# Clinical Decision Support for Pharmacogenomic Precision Medicine

# Clinical Decision Support for Pharmacogenomic Precision Medicine

## Foundations and Implementation

Edited by

**BETH DEVINE**

*Professor, The Comparative Health Outcomes, Policy and Economics (CHOICE) Institute, Shirley and Herb Bridge Endowed Professor for Women in Pharmacy; Adjunct Professor, Department of Biomedical Informatics, School of Medicine; Adjunct Professor, Department of Health Systems and Population Health, School of Public Health; Member, Institute for Public Health Genetics, School of Public Health; University of Washington, Seattle, WA, United States*

**RICHARD DAVID BOYCE**

*Associate Professor of Biomedical Informatics, University of Pittsburgh School of Medicine, Pittsburgh, PA, United States*

**KRISTIN WIISANEN**

*Clinical Professor, Department of Pharmacotherapy and Translational Research, Director, Graduate Program in Precision Medicine Associate Director, UF Health Precision Medicine Program; Associate Dean, Entrepreneurial Programs, University of Florida - College of Pharmacy, Gainesville, FL, United States*

ELSEVIER

ACADEMIC PRESS

An imprint of Elsevier

Academic Press is an imprint of Elsevier
125 London Wall, London EC2Y 5AS, United Kingdom
525 B Street, Suite 1650, San Diego, CA 92101, United States
50 Hampshire Street, 5th Floor, Cambridge, MA 02139, United States
The Boulevard, Langford Lane, Kidlington, Oxford OX5 1GB, United Kingdom



**Notices**
Knowledge and best practice in this field are constantly changing. As new research and experience broaden our understanding, changes in research methods, professional practices, or medical treatment may become necessary.

Practitioners and researchers must always rely on their own experience and knowledge in evaluating and using any information, methods, compounds, or experiments described herein. In using such information or methods they should be mindful of their own safety and the safety of others, including parties for whom they have a professional responsibility.

To the fullest extent of the law, neither the Publisher nor the authors, contributors, or editors, assume any liability for any injury and/or damage to persons or property as a matter of products liability, negligence or otherwise, or from any use or operation of any methods, products, instructions, or ideas contained in the material herein.

ISBN: 978-0-12-824453-1

For information on all Academic Press publications visit our website at https://www.elsevier.com/books-and-journals

*Publisher:* Stacy Masucci
*Acquisitions Editor:* Peter B. Linsley
*Editorial Project Manager:* Sara Valentino
*Production Project Manager:* Sreejith Viswanathan
*Cover Designer:* Miles Hitchen

Typeset by TNQ Technologies

# Contents

## SECTION 3 Adoption and delivery of clinical decision support for pharmacogenomic precision medicine

# Contributors

**Blake Atwood**
Sanford Health Imagenetics, Sanford Medical Genetics Laboratory, Sioux Falls, SD, United States; University of South Dakota Sanford School of Medicine, Sioux Falls, SD, United States

**Richard David Boyce**
Biomedical Informatics and Clinical and Translational Science Institute, Secondary Appointment in the School of Pharmacy, Secondary Appointment in the Intelligent Systems Program, School of Computing and Information, Biomedical Informatics Training Program, University of Pittsburgh, Pittsburgh, PA, United States

**Jhon Camacho**
Medical Informatics Solution, Architect I&E Meaningful Research, Bogotá, Colombia

**Beth Devine**
The Comparative Health Outcomes, Policy and Economics (CHOICE) Institute, School of Pharmacy; Department of Health Systems and Population Health, School of Public Health; Department of Biomedical Informatics, School of Medicine; Member, Institute for Public Health Genetics; Member, Plein Center for Geriatric Pharmacy Research, University of Washington, Seattle, WA, United States

**Henry Mark Dunnenberger**
Mark R. Neaman Center for Personalized Medicine NorthShore University Health System, Evanston, IL, United States

**Michele Erickson-Johnson**
Sanford Health Imagenetics, Sanford Medical Genetics Laboratory, Sioux Falls, SD, United States; University of South Dakota Sanford School of Medicine, Sioux Falls, SD, United States

**Debbie M. Figueroa**
Sanford Health Imagenetics, Sanford Medical Genetics Laboratory, Sioux Falls, SD, United States

**Allen J. Flynn**
Department of Learning Health Sciences, School of Information, University of Michigan, Ann Arbor, MI, United States

**Alison E. Fohner**
Department of Epidemiology and Institute of Public Health Genetics, University of Washington, Seattle, WA, United States

**Christine Formea**
Intermountain Healthcare, Intermountain Precision Genomics, Salt Lake City, UT, United States; Intermountain Department of Pharmacy, Intermountain Precision Genomics, Salt Lake City, UT, United States

**Bret Heale**
Intermountain Healthcare, Intermountain Precision Genomics, Salt Lake City, UT, United States

**Harry Hochheiser**
Department of Biomedical Informatics, University of Pittsburgh, Pittsburgh, PA, United States

**Ender Karaca**
Sanford Health Imagenetics, Sanford Medical Genetics Laboratory, Sioux Falls, SD, United States

**Shayna R. Killam**
Department of Biomedical and Pharmaceutical Sciences and Skaggs Institute for Health Innovation, University of Montana, Missoula, MT, United States

**Clarissa Klein**
Department of Biomedical Data Science, Stanford University, Stanford, CA, United States

**Teri E. Klein**
Department of Biomedical Data Science and Medicine (BMIR), Stanford University, Stanford, CA, United States

**Everett Lally**
Department of Cancer Genetics and Prevention, Seattle Cancer Care Alliance, Seattle, WA, United States

**Mercy Laurino**
Department of Cancer Genetics and Prevention, Seattle Cancer Care Alliance, Seattle, WA, United States

**Wayne Liang**
Emory University School of Medicine, Pediatric Bone Marrow Transplant Physician and Physician Informaticist, Children's Healthcare of Atlanta, Emory School of Medicine, Emory University, Atlanta, GA, United States

**Amanda Massmann**
Sanford Imagenetics, Sanford Health, Sioux Falls, SD, United States; Department of Internal Medicine, University of South Dakota Sanford School of Medicine, Sioux Falls, SD, United States

**Khoa Nguyen**
Pharmacotherapy and Translational Research, College of Pharmacy, University of Florida, Gainesville, FL, United States

**Katrina M. Romagnoli**
Department of Translational Data Science and Informatics, Geisinger, Danville, PA, United States

**April Schultz**
Sanford Imagenetics, Sanford Health, Sioux Falls, SD, United States; Department of Internal Medicine, University of South Dakota Sanford School of Medicine, Sioux Falls, SD, United States

**James M. Stevenson**
Department of Medicine, Division of Clinical Pharmacology, Johns Hopkins University School of Medicine, Baltimore, MD, United States

**Casey Overby Taylor**
Department of Medicine, Division of General Internal Medicine, Johns Hopkins University School of Medicine, Baltimore, MD, United States; Department of Biomedical Engineering, Johns Hopkins University, Baltimore, MD, United States; Institute for Computational Medicine, Johns Hopkins University, Baltimore, MD, United States

**Joel Van Heukelom**
Sanford Imagenetics, Sanford Health, Sioux Falls, SD, United States; Department of Internal Medicine, University of South Dakota Sanford School of Medicine, Sioux Falls, SD, United States

**Kyle G. Volk**
Department of History, University of Montana, Missoula, MT, United States

**Nephi Walton**
Intermountain Healthcare, Intermountain Precision Genomics, Salt Lake City, UT, United States

**Michelle Whirl-Carrillo**
Department of Biomedical Data Science, Stanford University, Stanford, CA, United States

**Kristin Wiisanen**
UF Health Precision Medicine Program, Center for Pharmacogenomics and Precision Medicine, Department of Pharmacotherapy and Translational Research, College of Pharmacy, University of Florida, Gainesville, FL, United States

**Erica L. Woodahl**
Department of Biomedical and Pharmaceutical Sciences and Skaggs Institute for Health Innovation, University of Montana, Missoula, MT, United States

# About the editors

Dr. Devine is a health services researcher and health economist specializing in medication safety and pharmacogenomics in the context of clinical informatics and electronic health records. She estimates the value of implementing clinical decision support systems for pharmacogenomics from a Learning Health System perspective. Her work is funded by the Agency for Healthcare Research and Quality. She is a co-investigator in the Clinical Sequencing Evidence-Generating Research 2 (CSER 2) and Electronic Medical Records and Genomics (eMERGE IV) research networks funded by the National Human Genome Research Institute. She received her PhD from University of Washington (UW) Department of Health Services, PharmD from University of the Pacific, MBA from University of San Francisco, and completed a postdoctoral fellowship in health economics and outcomes research with UW and Roche. She is on the board of directors for International Society for Health Economics and Outcomes Research and associate editor for *Value in Health.*

Dr. Boyce is Associate Professor of Biomedical Informatics at the University of Pittsburgh. The use of informatics to support safe and effective medication therapy for older adults has been Dr. Boyce's primary interest since the early stages of his research career. He has more than 15 years of research experience at the intersection of in informatics, clinical research, and pharmacoepidemiology. He has led research projects focusing on improving medication safety that were funded by the United States National Institutes of Health and the Agency for Healthcare Research and Quality. He currently leads the Informatics Core for the Center of Excellence for Natural Product—Drug Interaction Research funded by the National Center for Complementary and Integrative Health. His publication record includes more than 50 peer-reviewed journal articles at the intersection of medication safety, informatics, and clinical decision support.

Kristin Wiisanen is a Clinical Professor in the Department of Pharmacotherapy and Translational Research at the University of Florida (UF) College of Pharmacy. As Associate Director of the UF Health Precision Medicine Program, she leads implementation of clinical pharmacogenomics services across diverse patient care settings in UF Health. Her primary research focuses on factors influencing interdisciplinary adoption of clinical pharmacogenomics. She also leads development of innovative educational approaches for clinical pharmacogenomics and serves as Director of the Graduate Program in Precision Medicine, Director of Continuing Pharmacy Education, and Director of the Pharmacogenomics Residency and Clinical Fellowship programs. Her publication record includes more than 50 peer-reviewed journal articles on innovative clinical and educational approaches in pharmacogenomics and patient care. Dr. Wiisanen is a fellow of the American Pharmacists Association (APhA), the American College of Clinical Pharmacy, a UF Entrepreneurial Faculty Fellow, and Editor-in-Chief of APhA's flagship publication, *Pharmacy Today*.

SECTION ONE

# Foundations of clinical decision support and pharmacogenomic precision medicine

CHAPTER ONE

# Overview of effective pharmacogenomic clinical decision support

**Richard David Boyce[1], Jhon Camacho[2], Wayne Liang[3], Kristin Wiisanen[4], Beth Devine[5]**

[1]Biomedical Informatics and Clinical and Translational Science Institute, Secondary Appointment in the School of Pharmacy, Secondary Appointment in the Intelligent Systems Program, School of Computing and Information, Biomedical Informatics Training Program, University of Pittsburgh, Pittsburgh, PA, United States
[2]Medical Informatics Solution, Architect I&E Meaningful Research, Bogotá, Colombia
[3]Emory University School of Medicine, Pediatric Bone Marrow Transplant Physician and Physician Informaticist, Children's Healthcare of Atlanta, Emory School of Medicine, Emory University, Atlanta, GA, United States
[4]UF Health Precision Medicine Program, Center for Pharmacogenomics and Precision Medicine, Department of Pharmacotherapy and Translational Research, College of Pharmacy, University of Florida, Gainesville, FL, United States
[5]The Comparative Health Outcomes, Policy and Economics (CHOICE) Institute, School of Pharmacy; Department of Health Systems and Population Health, School of Public Health; Department of Biomedical Informatics, School of Medicine; Member, Institute for Public Health Genetics; Member, Plein Center for Geriatric Pharmacy Research, University of Washington, Seattle, WA, United States

## Contents



## Introduction

*Pharmacogenomics* is the field of study focused on how genomic variations among individuals affect their different responses to medications, with a particular focus on preventing adverse drug effects caused by genetic predisposition. It is an important subfield of precision (or personalized)

*Clinical Decision Support for Pharmacogenomic Precision Medicine*
ISBN: 978-0-12-824453-1
https://doi.org/10.1016/B978-0-12-824453-1.00010-5

medicine, the field of study that focuses on how individual variation in biology, environment, lifestyle, and preferences impacts the prevention, diagnosis, and treatment of diseases. The philosophy of precision medicine is to obtain a comprehensive description of an individual's specific traits and characteristics (i.e., a *phenotype*) using genomic and/or other advanced methods in order to provide the most appropriate medical intervention that optimizes benefit and minimizes negative consequences for the individual. This approach differs from conventional medicine that tends to provide medical interventions that, on average, benefit individuals within a population, without the use of phenotype data that predicts that an individual patient's likely response. Put simply, precision medicine aims to provide a patient with the most effective individualized treatment, while conventional medicine provides the patient with the treatment that is most effective on average. Box 1.1 provides a brief example of the precision medicine perspective on patient care.

**Box 1.1 Precision medicine clinical vignette**

A 60-year-old Asian woman presents to an emergency room with acute, severe chest pain. She underwent a medical workup and was diagnosed with an acute myocardial infarction due to a severe blockage in a major coronary artery. She underwent emergent cardiac catheterization to open up the blockage and was started on clopidogrel, a glycoprotein (GP) IIb/IIIa inhibitor with antiplatelet activity. A CYP1C19 pharmacogenomic panel test was ordered as part of the institutional clopidogrel clinical decision support order set. The patient was admitted to the hospital. Three days later, the treating physician received an electronic critical laboratory value alert. The pharmacogenomic panel test revealed that the patient has a homozygous *CYP2C19*2/*2* loss of function variant, resulting in a "poor metabolizer" phenotype associated with decreased effectiveness of clopidogrel in preventing major cardiac events following an acute myocardial infarction. The alert also provided guidelines-based recommendations for switching to an alternative antiplatelet agent unaffected by this variant. The patient was switched to an alternative antiplatelet agent and was discharged from the hospital. She continues to take this medication and has not experienced any additional cardiovascular events.

## Pharmacogenomics and precision medicine

In pharmacogenomic medicine, healthcare providers obtain a profile of a patient's pharmacogenomic variants (i.e., genomic variants relevant to their response to specific medications), in order to select the right pharmacological intervention predicted to maximize therapeutic benefit and minimize harm from lack of effectiveness or from excessive toxicity. Individuals who receive therapeutic benefit are termed "responders," while those who experience lack of effectiveness are termed "under- or nonresponders," and those who experience excessive toxicity are termed "toxic responders." Other subfields of precision medicine not covered in this textbook include *precision diagnostics* (i.e., obtaining and defining a more accurate or precise diagnosis of an individual's disease through advanced methods such as genomic testing), *targeted therapies* (i.e., developing and providing treatments tailored to an individual's biological profile), and *gene therapy* (i.e., altering a patient's genetic or genomic makeup for therapeutic effect).

Pharmacogenomic science is frequently applied to patient care through clinical guidelines. First, new pharmacogenomic variant knowledge is discovered through basic genomic research. Next, newly discovered pharmacogenomic variants are reported in the scientific literature and in variant databases. Expert panels then evaluate the scientific evidence and produce clinical guidelines to guide clinical decision-making. Collaboratives such as the Clinical Pharmacogenetics Implementation Consortium ([CPIC] (Relling & Klein, 2011); and the Dutch Pharmacogenetics Working Group (DPWG) are leaders in publishing openly available, peer-reviewed, and evidence-based pharmacogenomic clinical practice guidelines. These guidelines may then be adopted by healthcare providers and health provider organizations into clinical practice. Patient outcomes data may be collected and analyzed to generate new pharmacogenomic knowledge within the genomic learning healthcare system, or one in which clinical practice and research influence each other with the goal of improving the efficiency and effectiveness of disease prevention, diagnosis, and treatment (Genomics-Enabled Learning Health Care Systems: Gathering and Using Genomic Information to Improve Patient Care and Research: Workshop Summary, 2015). In general, a learning health system (LHS) approach is where a community forms and commits to learn and then apply new insights continuously. As Fig. 1.1 shows, a genomic learning healthcare system seeks to learn and apply new insights on how to best apply genomic information to arrive at the best clinical outcomes possible (Green et al., 2020).

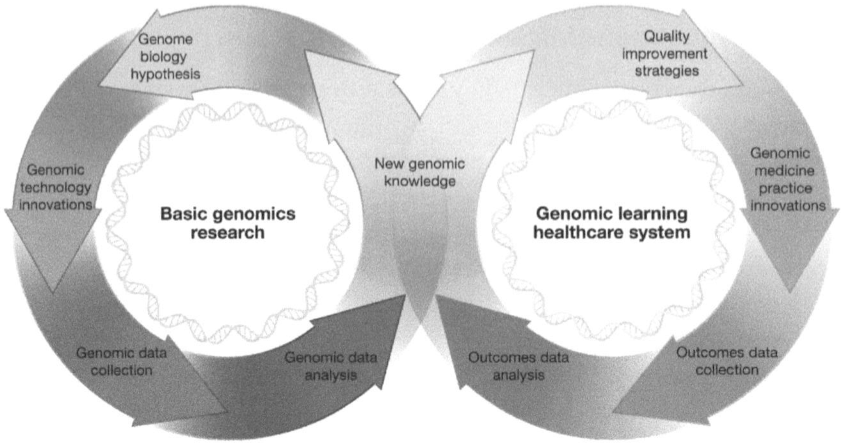


**Figure 1.1** Genomic research and learning healthcare system virtual cycles. Basic genomic research informs genomic clinical care, and clinical care informs research. *From Green, E. D., Gunter, C., Biesecker, L. G., Di Francesco, V., Easter, C. L., Feingold, E. A., Felsenfeld, A. L., Kaufman, D. J., Ostrander, E. A., Pavan, W. J., Phillippy, A. M., Wise, A. L., Dayal, J. G., Kish, B. J., Mandich, A., Wellington, C. R., Wetterstrand, K. A., Bates, S. A., Leja, D., … Manolio, T. A. (2020). Strategic vision for improving human health at The Forefront of Genomics.* Nature, *586(7831), 683–692. https://doi.org/10.1038/s41586-020-2817-4.*

Pharmacogenomic knowledge is generated through basic genomic research, and is integrated into clinical care through practice innovations such as pharmacogenomic computerized clinical decision support (CDS) within electronic health record (EHR).

Despite the availability of clinical guidelines, pharmacogenomics is not yet widely adopted within mainstream clinical practice, preventing many patients from fully benefitting from pharmacogenomic medicine. There are several factors that have slowed the adoption of pharmacogenomics. Healthcare payers have been slow to add reimbursement for pharmacogenomic testing and consultation because they require that such procedures have demonstrated analytical and clinical validity as well as clinical utility. Complicated reimbursement policies make it difficult for providers to know which of the dozens of pharmacogenomics tests will be covered by insurance. Prescribers tend to have a favorable perception of the potential value of pharmacogenomics testing (Amara et al., 2018; Deininger et al., 2020; Jameson et al., 2021). However, they express concerns about how to best apply the results to improve care, uncertainty about which sources of clinical pharmacogenomics knowledge are trustworthy, and concerns about the potential negative financial impact on patients. While pharmacists can help provide prescribing guidance specific to clinical phenotypes, pharmacogenomics training has only recently been added to pharmacy curriculums and many pharmacists find it hard to find relevant information in drug product labeling (Romagnoli et al., 2016).

## Potential benefits of CDS

A *clinical decision support (CDS) system* is a computer program that is actively involved in the clinical decision-making process by delivering patient-specific recommendations to facilitate high-quality patient care. As computer programs, CDS systems are distinct from noncomputerized clinical decision aids. Therefore, clinical practice guidelines and other documents (even in digital format) are not considered CDS. A CDS system is active in the decision-making process of the clinician for each patient, as opposed to passively presenting normative information. Systems that fall outside this criterion could include a static flowchart depicting a decision algorithm or a mobile application that lists a series of recommendations but does not filter them (or otherwise process them) depending on each patient's condition. This does not mean that the CDS system needs to interact, call attention, or interrupt the health provider every time a decision needs to be made. Instead, a least-disruptive CDS should process a patient's health information in the background and present recommendations only when warranted.

CDS has the potential to address several barriers to broaden pharmacogenomics adoption. It could help to fill in clinicians' knowledge gaps about pharmacogenomic phenotypes and enable them to apply the most relevant care guidelines for each specific patient. The widespread adoption of EHRs that are capable of providing patient phenotype data is important for providing clinicians with contextualized and actionable pharmacogenomics alerts. It is also possible that the clinical actions suggested by a CDS tool could be linked to current information about reimbursement, thereby addressing cost concerns. Pharmacogenomic CDS could be directed at prescribers, to assist with decision-making during ordering, and also pharmacists, as another layer of protection for patients when prescription orders are filled and dispensed.

CDS involves delivery of patient-specific recommendations to facilitate high-quality patient care. Patient-specific recommendations mean that CDS's output needs to apply to the patient receiving care at the time. This requirement excludes systems that present reminders (or suggest materials to review) that apply to a population but not necessarily to the patient at hand. A result of this criterion is that the value a CDS system adds to a clinical process is directly related to the relevance of the advice it provides. Therefore, successful CDS is expected to deliver recommendations that apply to the current patient and the specific decision being made. This does not mean that a CDS has to make itself noticed every time a decision is made, but should provide advice only when it is relevant. In other words, it should interrupt the health provider decision process only when such interruption has a reasonable likelihood of improving patient care. How to estimate that probability, set the threshold, and provide advice in a least-disruptive manner are major challenges in CDS design.

A CDS system can be *stand-alone* or *integrated* within other systems, and *passive* (requiring user retrieval) or *active* (automatically triggered without requiring user retrieval). CDS can be delivered in a variety of formats, including reference materials, data reports, electronic alerts, order sets, documentation templates, and others. Whenever possible, a CDS system is integrated with other information systems such as an electronic health records (EHR) and computerized physician order entry (CPOE). Such an integration carries two significant benefits. First, it reduces the disruption of the clinical process, allowing the CDS to operate without user intervention. Autonomous CDS operation means that the system does not necessarily add new work to the health providers. Second, integration improves the quality of the input data, reducing the likelihood of producing incorrect recommendations. An important requirement of CDS system is that it provides clinically relevant information to the clinician at the right location in the workflow of clinicians. The CDS service must run efficiently to meet clinician expectations and not interrupt from the clinical workflow. Delays in presenting CDS information may lead to unsuccessful CDS adoption and clinician frustration.

A CDS system disseminates and implements guidelines and protocols that affect clinical processes directly. It operates "just-in-time," i.e., at the time where clinical recommendations are most needed. Moreover, well-designed CDS systems implement evidence-based recommendations, which implies the formalization of unstructured narrative statements into computable decision support logic. This addresses a common criticism of clinical guideline documents as being ambiguous, unclear, and complex (Lee et al., 2015). By helping clinicians to adhere to best practice clinical guidelines, a CDS system can reduce unnecessary orders and discourage ineffective treatments (Sutton et al., 2020). It might also reduce provider workload by automating tasks involved in decision-making, such as searching the patient's EHR data for conditions and treatments that have a bearing on potential outcomes. Such a reduction in provider workload is predicated on a CDS design that maximizes the relevance of CDS-provided information and minimizes workflow disruption. The improvement in clinical decision-making achieved with CDS can result in several benefits for patients including a reduction in medical errors and adverse events (Helmons et al., 2015). Additionally, CDS can improve timeliness in treating diagnoses requiring urgent attention, reducing the likelihood of complications or death (Clinical Decision Support Decreases Sepsis Mortality in AL, 2016).

## Role of CDS in pharmacogenomic precision medicine

The National Institutes of Health (NIH) National Human Genome Research Institute (NHGRI) identifies CDS as the key method for integrating genomic knowledge into medical practice and making it readily accessible to more patients (Green et al., 2020). Within pharmacogenomics, CDS tools are akin to Global Positioning System (GPS) navigation tools that provide automobile drivers with turn-by-turn driving directions. Given that many providers lack expertise in genomics, including pharmacogenomics, the implementation of pharmacogenomic CDS tools has the potential to significantly impact pharmacogenomics-related outcomes in patients receiving care at those institutions. Additionally, through the integration of patient-specific, evidence-based alerts, pharmacogenomic CDS tools have the added potential role of promoting equity and inclusion among populations of patients who historically have benefitted the least from genetic testing due to factors such as age, ancestry, and gender.

## Developing and implementing CDS for pharmacogenomic precision medicine

Components of a successful implementation of pharmacogenomic CDS include both the clinical content and implementation factors. Clinical content refers to the scientific basis or medical rationale for the CDS, and may be based on an existing clinical guideline or other sources of scientific evidence. Local content area experts, such as physicians or pharmacists, generally lead and govern the clinical content for each CDS tool. Implementation factors refer to how the decision support is operationalized and delivered within the healthcare provider organization. The content area experts often initiate the pharmacogenomic CDS implementation project by submitting a new project request to the health information technology (IT) group, in order to achieve certain clinical objectives (e.g., avoid major cardiovascular events following acute myocardial infarction due to decreased effectiveness of clopidogrel). After relevant stakeholders have agreed to proceed with the project, a project team is convened to design and implement the pharmacogenomic CDS intervention. The implementation plan is usually developed collaboratively among relevant stakeholders, including content area experts, clinical informaticists, and information technology professionals, and implementation begins. After implementation, the same or a different group may be tasked with evaluating the tool's ongoing performance, to determine whether to continue the tool as is, or to make modifications to the tool, or to discontinue the CDS tool.

As with other practice-changing interventions, the successful implementation of pharmacogenomic CDS is a multidimensional challenge (Camacho et al., 2020). To reap the full benefits of a pharmacogenomic CDS system, implementers need to pay attention to factors internal and external to the health organization (Damschroder et al., 2009) and those related to the providers' preferences and their mental model about their practice (Box 1.2) (Cane et al., 2012).

## Box 1.2 Questions to guide a successful implementation of a pharmacogenomic CDS system

Implementers need to work closely with all relevant stakeholders to arrive at a common set of expectations for the pharmacogenomic CDS system and an implementation strategy that is likely to succeed. At the minimum, the answers to the following questions need to be clear in the mind of all personnel involved or affected by the implementation:

What is the objective of implementing the pharmacogenomic CDS?
What value will the pharmacogenomic CDS add to the clinical processes?
What value will the pharmacogenomic CDS add to my practice?
How will the clinical processes change?
How will my responsibilities change?
What is the decision algorithm behind the pharmacogenomic CDS?
What is the clinical evidence supporting the decision algorithm?
What steps will comprise the implementation strategy?
What measures will be taken to minimize disruption of clinical processes?
What measures will be taken to protect patient care/safety?

The CDS Five Rights model (Improving Medication Use and Outcomes with Clinical Decision Support: A Step-by-Step Guide, 2015) is a useful conceptual framework for designing a pharmacogenomic CDS implementation. At the beginning stages of designing a new pharmacogenomic CDS intervention, project teams can use the CDS Five Rights framework as a way to conceptualize decisions on how the intervention will be operationalized within the institution. After implementation, the CDS Five Rights framework can be used as one tool to evaluate whether the pharmacogenomic CDS intervention meets the requirements inspired by the Five Rights, or whether some design modification is required. Table 1.1 lists the CDS Five Rights applied generally to pharmacogenomic CDS along with data integration challenges that also need to be considered (Freimuth et al., 2017), while Table 1.2 provides an example of how the CDS Five Rights might be applied to a clopidogrel pharmacogenomic CDS tool, as referenced in Box 1.1.

In addition to assessing if the design of the pharmacogenomic CDS system satisfies the Five Rights, the pharmacogenomic CDS tool may be evaluated based on usage metrics, usability, user satisfaction, efficiency, and other measures (Measure Effects and Refine CDS Interventions, 2011). The impact of the system on clinical care may be evaluated using outcome measures or process measures. *Outcome measures* evaluate the key clinical or business outcomes of interest (e.g., rates of major cardiovascular events following myocardial infarction after clopidogrel initiation). *Process measures* evaluate key processes that may be related to achieving the outcomes of interest (e.g., percentage of patients who received appropriate pharmacogenomic testing at clopidogrel initiation; percentage of patients whose antiplatelet regimen was modified according to guidelines after an abnormal pharmacogenomic result). In general, outcome measures are preferred but are more difficult to obtain compared with process measures.

## Other considerations for pharmacogenomic CDS

Economic and financial factors are important in the implementation of pharmacogenomic CDS. Healthcare provider organizations often operate within tight margins with limited potential streams of revenue besides reimbursement for patient care. Information technology (IT) is often a major expense category that, although essential for operation and does not directly generate revenue. Therefore, healthcare provider organizations must carefully evaluate the return on investment on any new IT-related project,

**Table 1.1** CDS five rights framework applied to pharmacogenomics.

| CDS "right" | Principal points and example | Genomic data integration challenges |
|---|---|---|
| Right information | • Evidence-based information that may be derived from a guideline or national performance measure<br>• Constrained to relevant clinical information to avoid cognitive overload<br>• Supported by links to educational material (e.g., Infobuttons)<br>• Targets clinicians and patients | • Lack of standardized and structured data for genomic results, clinical observations, and clinical phenotypes<br>• Gaps in knowledge of relevant clinical factors<br>• Lack of an accepted minimal information model for genomic medicine<br>• Few rule definitions that are portable across clinical sites<br>• A need for scalable methods to maintain data, knowledge, and rules |
| Right person | • An individual who can act on the information<br>• Might be a clinician, patient, or caregiver | • Localized clinical workflows and team structures limit generalizability of interventions<br>• Information must be tailored for each user based on their needs (and kept in sync) |
| Right intervention format | • Fits the problem that the intervention is trying to address<br>• May be active (e.g., interruptive alerts) or passive<br>• Examples include alerts, order sets, Infobuttons, dosing calculators, protocols, and decision trees | • Lack of evidence comparing different types of interventions<br>• Existing clinical systems (e.g., EHRs) and workflows do not fully support integration of genomic data |
| Right channel | • Provider-facing clinical systems, such as the EHR or custom applications (e.g., dashboards) | • Lack of user-friendly tools to deliver complex information<br>• Need standard interfaces that enable the integration of tools and methods for |

**Table 1.1** CDS five rights framework applied to pharmacogenomics.—cont'd

| CDS "right" | Principal points and example | Genomic data integration challenges |
|---|---|---|
| | • Patient-facing systems (e.g., portal or personal health record)<br>• Hybrid systems, such as mobile apps that facilitate provider-patient communication or collect personalized health data (e.g., activity tracker, glucose monitor) | delivering genomic information into customized clinical systems |
| Right time in workflow | • Information delivered precisely when it will have the best impact on decision-making<br>• Requires a thorough understanding of clinical workflows<br>• Different care settings may make the best use of information at different times within a given workflow | • Genomic data may not be available (e.g., from preemptive testing), test turnaround time usually exceeds clinical decision-making time |

CDS can help improve health outcomes if the Right Information is communicated to the Right Person using the Right Intervention Format delivered through the Right Channel at the Right Time in the Workflow. Some challenges to achieving the "five rights" for genomic data through its integration into clinical systems and workflows are listed in the right-hand column. CDS, clinical decision support; EHR, electronic health record.

including pharmacogenomic CDS implementation projects. Matthias et al. estimated that the cost of developing a new genomic CDS alert within an institution is $4600 (Mathias et al., 2016). Each CDS rule should be compared with the potential cost savings and clinical benefit of each alert, and costs may be reduced significantly through increased cooperation in CDS development and technical interoperability across institutions.

Genetic counseling is an important service that should be available to patients whenever any genomic testing is performed. Pharmacogenomic testing is a form of genomic testing, so genetic counseling should be offered at the time of testing. An individual's genomic sequence is highly sensitive, as it is highly unique, largely unaltered throughout the entire lifespan, fully identifiable, heritable, and diagnostic of human diseases. Genomic findings

**Table 1.2** CDS five rights applied to clopidogrel pharmacogenomics.

| CDS five rights | Sample Clopidogrel Pharmacogenomics Clinical Decision Support (CDS) Tool |
|---|---|
| The Right Information | Institution-specific recommendations, based on clinical guidelines. |
| The right person | Order set: The **prescribing provider** for clopidogrel. Critical result flag: **All users** reviewing the laboratory tab. Critical result notification: The **prescribing provider** and the **pharmacist group**. |
| The right format | 1. **Order set**. Mandatory for users initiating clopidogrel as a new medication. The clopidogrel pharmacogenomic panel test is included as an orderable that is selected by default. A concise statement providing the scientific rationale is included. 2. **Critical result flag**. Any clinically relevant clopidogrel pharmacogenomic result is flagged as a critical result within the EHR, visible to all users. 3. **Critical result notification**. An electronic message is sent to the ordering provider and the pharmacist group indicating the abnormal result with recommended clinical action (e.g., change dosing, monitoring considerations, change agent). |
| The right channel | Within the EHR. |
| The right time in the workflow | Order set: At initiation of clopidogrel. Critical result flag: At laboratory results review. Critical result notification: When laboratory has processed the pharmacogenomic lab results. |

The CDS Five Rights model can be applied to specific medications, such as clopidogrel, which is affected by variability in the CYP2C19 gene.

have the potential to impact patients and their families medically, psychologically, and socially for the rest of their lives. For example, genomic testing may uncover misattributed parentage, discordance between genotypic and phenotypic sex, risk of severe disease or medical conditions, and others. Genomic findings might affect the individual's ability to obtain life insurance or put them at risk for other forms of genomic discrimination. Once genomic information is obtained and disclosed, disclosure cannot be reverse. Even genetic testing obtained in the newborn and childhood period can have lifelong impact. Highly focused pharmacogenomic testing sidesteps some of these challenges, as findings are use not to diagnose medical conditions, but to prevent adverse effects of pharmaceuticals. However, given the potential risks as described above, patients should be counseled on the risks and benefits of pharmacogenomic testing before it is obtained.

## Overview of this book

The goal of this book is to inform learning health system stakeholders, including clinicians and informaticists, about implementing CDS for precision medicine. This includes the knowledge foundation of pharmacogenomic CDS, how pharmacogenomic data gets from the laboratory into electronic health records, current methods and infrastructure for CDS, and key considerations for implementation and administration including equity, access, and patient counseling aspects. Fig. 1.2 provides an overview of the book's organization which is arranged into three sections.

The first section, which includes this chapter and three others, discusses foundational concepts needed for successful pharmacogenomic CDS, including the relevant sources of scientific evidence and knowledge, laboratory considerations, and the important need for equity and diversity within all endeavors. Anyone reading this book who is new to implementing CDS or pharmacogenomics will benefit from reading these chapters.

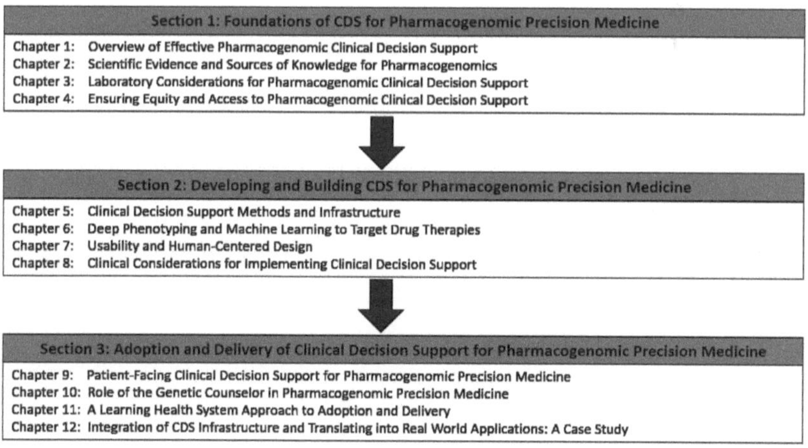


**Figure 1.2** Overview of clinical decision support for pharmacogenomic precision medicine. This text is divided into three sections. Although each chapter may be beneficial as a standalone reading, when approaching CDS implementation from start to finish, the sections and flow of the chapters can be seen as building blocks, with chapters in each section building on the previous one.

The chapters in Section 2 go into further detail about developing and building CDS for pharmacogenomic precision medicine. The first three chapters of this section focus on the more technical details of pharmacogenomic CDS including methods and infrastructure, deep phenotyping and the role of machine learning, and how to assess the usability of pharmacogenomic CDS systems. The final chapter in this section focuses on clinical considerations when building and implementing CDS. This seminal chapter explores further the five rights of CDS and provides insight on essential clinical considerations based on authors' experiences. Persons who are planning to work on the technical design and build of a pharmacogenomic CDS system are encouraged to read these chapters.

The final section of this book focuses on successful adoption, delivery, and optimization of CDS for pharmacogenomic precision medicine. The first two chapters focus on patient-facing CDS and the role of the genetic counselor, while the third details the importance of a learning health system approach to building and maintaining pharmacogenomic CDS systems as well as influencing clinical considerations. The final chapter provides a longitudinal case study that ties all of the topics discussed in the other chapters of the book together and provides real-world examples of issues that may arise in developing and implementing CDS for pharmacogenomic precision medicine.

## Conclusion

Pharmacogenomics is a subfield of precision medicine that focuses on maximizing therapeutic benefit (responders) while minimizing harm (underresponders and toxic responders) due to individual biological variability in response to pharmaceuticals. Scientific discoveries in pharmacogenomic science are translated into clinical guidelines, which can then be integrated into clinical practice through pharmacogenomic CDS. The design, implementation, and maintenance of successful pharmacogenomic CDS involves many stakeholders and requires consideration of several topics ranging from knowledge sources to CDS methods to clinical and counseling considerations. This book provides detailed discussions of these topics so that stakeholders, including clinicians and informaticists, can achieve the many benefits of successful pharmacogenomic CDS for precision medicine.

## References

Amara, N., Blouin-Bougie, J., Bouthillier, D., & Simard, J. (2018). On the readiness of physicians for pharmacogenomics testing: An empirical assessment. *The Pharmacogenomics Journal, 18*(2), 308–318. https://doi.org/10.1038/tpj.2017.22

Camacho, J., Zanoletti-Mannello, M., Landis-Lewis, Z., Kane-Gill, S. L., & Boyce, R. D. (2020). A conceptual framework to study the implementation of clinical decision support systems (BEAR): Literature review and concept mapping. *Journal of Medical Internet Research, 22*(8), e18388. https://doi.org/10.2196/18388

Cane, J., O'Connor, D., & Michie, S. (2012). Validation of the theoretical domains framework for use in behaviour change and implementation research. *Implementation Science : IS, 7*, 37. https://doi.org/10.1186/1748-5908-7-37

*Clinical decision support Decreases Sepsis mortality in AL.*(2016). https://ehrintelligence.com/news/clinical-decision-support-decreases-sepsis-mortality-in-al.

Damschroder, L. J., Aron, D. C., Keith, R. E., Kirsh, S. R., Alexander, J. A., & Lowery, J. C. (2009). Fostering implementation of health services research findings into practice: A consolidated framework for advancing implementation science. *Implementation Science : IS, 4*, 50. https://doi.org/10.1186/1748-5908-4-50

Deininger, K. M., Tsunoda, S. M., Hirsch, J. D., Anderson, H., Lee, Y. M., McIlvennan, C. K., Page, R. L., Tran, J. N., & Aquilante, C. L. (2020). National survey of physicians' perspectives on pharmacogenetic testing in solid organ transplantation. *Clinical Transplantation, 34*(10), e14037. https://doi.org/10.1111/ctr.14037

Freimuth, R. R., Formea, C. M., Hoffman, J. M., Matey, E., Peterson, J. F., & Boyce, R. D. (2017). Implementing genomic clinical decision support for drug-based precision medicine. *CPT: Pharmacometrics & Systems Pharmacology, 6*(3), 153–155. https://doi.org/10.1002/psp4.12173

*Genomics-enabled learning health care systems: Gathering and using genomic information to improve patient care and research: Workshop summary.*(2015). https://doi.org/10.17226/21707

Green, E. D., Gunter, C., Biesecker, L. G., Di Francesco, V., Easter, C. L., Feingold, E. A., Felsenfeld, A. L., Kaufman, D. J., Ostrander, E. A., Pavan, W. J., Phillippy, A. M., Wise, A. L., Dayal, J. G., Kish, B. J., Mandich, A., Wellington, C. R., Wetterstrand, K. A., Bates, S. A., Leja, D., … Manolio, T. A. (2020). Strategic vision for improving human health at the Forefront of Genomics. *Nature, 586*(7831), 683–692. https://doi.org/10.1038/s41586-020-2817-4

Helmons, P. J., Suijkerbuijk, B. O., Nannan Panday, P. V., & Kosterink, J. G. W. (2015). Drug-drug interaction checking assisted by clinical decision support: A return on investment analysis. *Journal of the American Medical Informatics Association : JAMIA, 22*(4), 764–772. https://doi.org/10.1093/jamia/ocu010

*Improving medication use and outcomes with clinical decision support: A step-by-step guide.*(2015). https://www.ismp.org/resources/improving-medication-use-and-outcomes-clinical-decision-support-step-step-guide.

Jameson, A., Fylan, B., Bristow, G. C., Sagoo, G. S., Dalton, C., Cardno, A., Sohal, J., & McLean, S. L. (2021). What are the barriers and enablers to the implementation of pharmacogenetic testing in mental health care settings? *Frontiers in Genetics, 12*, 740216. https://doi.org/10.3389/fgene.2021.740216

Lee, P. Y., Liew, S. M., Abdullah, A., Abdullah, N., Ng, C. J., Hanafi, N. S., Chia, Y. C., Lai, P. S. M., Wong, S. S. L., & Khoo, E. M. (2015). Healthcare professionals' and policy makers' views on implementing a clinical practice guideline of hypertension management: A qualitative study. *PloS One, 10*(5), e0126191. https://doi.org/10.1371/journal.pone.0126191

Mathias, P. C., Tarczy-Hornoch, P., & Shirts, B. H. (2016). Modeling the costs of clinical decision support for genomic precision medicine. *American Medical Informatics Association Joint Summits on Translational Science Proceedings. American Medical Informations Association Joint Summits on Translational Science, 2016*, 60–64.

*Measure effects and refine CDS interventions.*(2011). https://www.healthit.gov/sites/default/files/3-4-5-measure-effects-and-refine-cds-interv.pdf.

Relling, M. V., & Klein, T. E. (2011). CPIC: Clinical Pharmacogenetics implementation Consortium of the pharmacogenomics research network. *Clinical Pharmacology and Therapeutics, 89*(3), 464–467. https://doi.org/10.1038/clpt.2010.279

Romagnoli, K. M., Boyce, R. D., Empey, P. E., Adams, S., & Hochheiser, H. (2016). Bringing clinical pharmacogenomics information to pharmacists: A qualitative study of information needs and resource requirements. *International Journal of Medical Informatics, 86*, 54—61. https://doi.org/10.1016/j.ijmedinf.2015.11.015

Sutton, R. T., Pincock, D., Baumgart, D. C., Sadowski, D. C., Fedorak, R. N., & Kroeker, K. I. (2020). An overview of clinical decision support systems: Benefits, risks, and strategies for success. *NPJ Digital Medicine, 3*, 17. https://doi.org/10.1038/s41746-020-0221-y

CHAPTER TWO

# Scientific evidence and sources of knowledge for pharmacogenomics

**Michelle Whirl-Carrillo[1], Clarissa Klein[1], Teri E. Klein[2]**
[1]Department of Biomedical Data Science, Stanford University, Stanford, CA, United States
[2]Department of Biomedical Data Science and Medicine (BMIR), Stanford University, Stanford, CA, United States

## Contents



## Introduction

This chapter discusses and reviews sources of scientific evidence and centralized resources that can be used to support clinical implementation

*Clinical Decision Support for Pharmacogenomic Precision Medicine*
ISBN: 978-0-12-824453-1
https://doi.org/10.1016/B978-0-12-824453-1.00012-9

of pharmacogenomic knowledge. The most widely used source of evidence in pharmacogenomics is peer-reviewed published literature, but there are also unreviewed studies such as those in bioRxiv (https://www.biorxiv.org/), an online archive of preprints and unpublished data. While the latter sources are more difficult to utilize as they frequently have not been vetted, some authoritative agencies such as the US Food and Drug Administration (FDA) use unpublished information from premarket studies in drug labeling. These must follow certain guidelines that the FDA has developed and so can serve as trusted sources of unpublished information. Publicly available online resources discussed in this section include those that catalog and synthesize information from these primary evidence sources (The Pharmacogenomics Knowledgebase [PharmGKB], the Clinical Pharmacogenetics Implementation Consortium [CPIC], and the Dutch Pharmacogenetics Working Group [DPWG]) or receive primary data submitted directly to them (The Pharmacogene Variation Consortium [PharmVar]). Implementation tools (Pharmacogenomics Clinical Annotation Tool [PharmCAT]) also build on top of these resources.

## Sources of scientific evidence of pharmacogenomic associations

Currently, various sources of scientific evidence exist to support clinical implementation of pharmacogenomic knowledge, ranging from databases of peer-reviewed literature to the inclusion of pharmacogenomic information in drug labels, to databases that specifically catalog and curate pharmacogenomic knowledge.

Primary scientific evidence supporting pharmacogenomic associations is most widely curated from peer-reviewed published literature. The National Institutes of Health (NIH)-funded PubMed (https://pubmed.ncbi.nlm.nih.gov) is a publicly accessible search engine that indexes over 32 million citations within the NIH's MEDLINE database. Pharmacogenomic literature available within PubMed covers a wide range of study types, from clinical studies to in vitro assays. While randomized controlled trials (RCTs) are a gold standard in the medical community, there are exceptionally few RCTs in pharmacogenomics. Studying low-frequency alleles necessitates exceptionally large cohorts, and there are ethical issues in potentially improperly dosing a group of subjects, both of which make this format less feasible for widespread pharmacogenomic study (Huddart, Sangkuhl, et al., 2019) Although the potential value of such studies and the rationale for their relative absence is debatable, the fact remains that pharmacogenomic RCTs are few and far between.

Other types of pharmacogenomic clinical studies that can support the association between genetic variants and clinical outcomes, and that are more commonly found, include observational cohort or case-control studies, case reports or case series, and metaanalyses. Pharmacogenomic studies are commonly hypothesis-driven and frequently focused on candidate genes and variants that are already suspected to have some impact on drug response based on a mechanistic understanding of the drug's pharmacokinetics (PK) or pharmacodynamics (PD). Genome-wide association studies (GWAS) are also used to discover novel variants across the genome that correlate with a particular outcome, and typically require subsequent studies for association verification and to determine causation. In addition to clinical outcome studies, metabolism and function studies can make valuable contributions to the evidence for a pharmacogenomic association. Studies of drug clearance in patients, enzymatic activity in liver microsomes, and functional in vitro assays add to the wider body of knowledge.

In addition to peer-reviewed literature, the FDA (http://www.fda.gov) serves as another resource for evidence supporting pharmacogenomic associations. The FDA approves therapeutic product labeling, maintains a list of products that contain genomic biomarkers on their label, and periodically updates their "Table of Pharmacogenomics Biomarkers in Drug Labeling" (https://www.fda.gov/drugs/science-and-research-drugs/table-pharmacogenomic-biomarkers-drug-labeling). According to the FDA's website, biomarkers include both germline and somatic gene variants and exclude nonhuman genetic variants such as microbial variants that are targeted by antibiotics. Biomarkers used to diagnose disease are included on the FDA's Pharmacogenomics Biomarker list if they are also somehow connected to the drug activity in general or in a specific subpopulation. The labels themselves can include guidance on whether to conduct or consider conducting a pharmacogenomic test before prescribing and how prescribing might be altered based on genetic test results, including situations in which the drug would be contraindicated. Many labels reference genes involved in drug exposure, mechanism of action, or trial design rather than testing or genotype-based prescribing. These biomarkers may be mentioned in any section of their corresponding drug label, and there is no requirement for the label to cite published studies as supporting evidence.

The FDA has published guidelines for evaluating pharmacogenomics in early-phase clinical studies as well as recommendations for reporting on this pharmacogenomic data in drug labeling (*Clinical Pharmacogenomics: Premarket Evaluation in Early-Phase Clinical Studies and Recommendations for Labeling*, 2013). This includes both the general recommendation that pharmacogenomic information should be included "only if it is useful to inform prescribers about the impact (or lack of impact) of genotype on phenotype," or, if testing for a variant is available, whether this testing should be considered, is recommended, or is necessary. While FDA labels are not required to cite published studies as supporting evidence in support of biomarkers, the guidance for evaluating pharmacogenomics states that data supporting clinical recommendations based on pharmacogenomics should be included "when the information has important implications for the safe and effective use of a drug" (e.g., recommendations for restricted use, dosage, contraindications).

As of 2020, the FDA maintains a separate "Table of Pharmacogenetic Associations" (https://www.fda.gov/medical-devices/precision-medicine/table-pharmacogenetic-associations) that lists gene–drug pairs that the FDA "has evaluated and believes there is sufficient evidence to suggest that subgroups of patients with certain genetic variants, or genetic variant-inferred phenotypes … are likely to have altered drug metabolism, and in certain cases, differential therapeutic effects, including differences in risks of adverse events." It is not stated what evidence was evaluated by the FDA for the entries in this table, and no citations are provided. While there is some overlap between the two tables, and some of the biomarkers from the "Table of Pharmacogenomic Biomarkers in Drug Labeling" are included in the "Table of Pharmacogenetic Associations," most are not, and the overlap is not extensive. Similarly, genes are listed on the "Table of Pharmacogenetics Associations" that are not included in the "Table of Pharmacogenomic Biomarkers in Drug Labeling."

## Sources of pharmacogenomic knowledge

In addition to peer-reviewed primary evidence in the form of literature and the FDA's inclusion of pharmacogenomics in their drug labeling, there are several public resources designed specifically for cataloging and curating pharmacogenomic knowledge, including the PharmVar (Gaedigk et al., 2018, 2019), PharmGKB (Whirl-Carrillo et al., 2012, 2021), and CPIC (Relling et al., 2019; Relling & Klein, 2011) projects. While the three projects are distinct in structure, purpose, and focus, they collectively utilize the knowledge provided by the others, maintain working relationships, and occasionally collaborate on projects. PharmVar catalogs pharmacogene

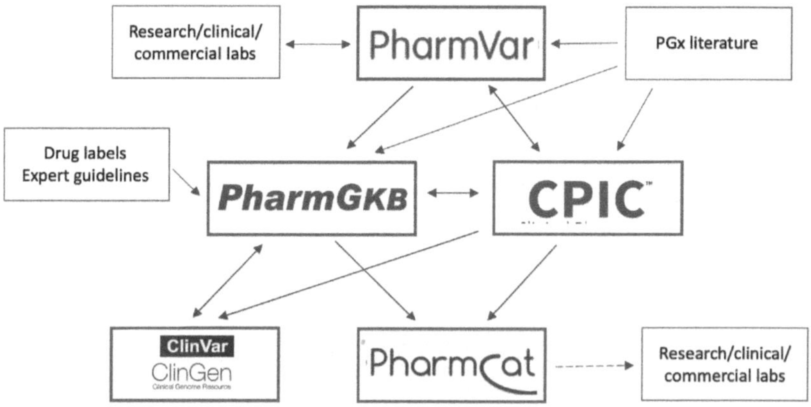


**Figure 2.1** Relationships among prominent, publicly available pharmacogenomic resources. Multiple resources work collaboratively and independently to provide evidence-based pharmacogenomic data sources.

alleles and maintains a nomenclature for them. This allele nomenclature is then used directly by both PharmGKB and CPIC. PharmGKB curates pharmacogenomic information from various sources (including PharmVar, CPIC, DPWG), annotates the curated information, synthesizes multiple pieces of annotated information to generate newly integrated knowledge, and disseminates this integrated pharmacogenomic information in various formats through its website. Several types of PharmGKB annotations are used by CPIC to evaluate gene–drug pairs for potential guideline creation. CPIC reviews pharmacogenomic literature to provide clinicians with peer-reviewed, updated, published guidelines with genotype-based drug prescribing recommendations. These guidelines are then annotated in PharmGKB, and CPIC's allele function assignments (included in the guidelines) are displayed on the PharmVar website. On an implementation level, the clinical annotation tool PharmCAT combines information from these three resources and annotates patient genotypes with prescribing recommendations. The following sections describe the history of and various relationships among these resources, which are illustrated in Fig. 2.1 below.

## Pharmacogene variation consortium

The Pharmacogene Variation Consortium (PharmVar; https://www.pharmvar.org) is the primary allelic nomenclature resource for multiple pharmacogenes, including the cytochrome P450 genes, and serves as a centralized pharmacogene variation data repository (Gaedigk et al., 2018, 2019). PharmVar extensively evaluates and catalogs pharmacogenomic alleles using the star (*) allele nomenclature, which is used broadly across pharmacogenomics publications, testing platforms, clinical labs, and clinical test results.

Many clinicians and scientists outside of the pharmacogenomics community are unfamiliar with star (*) alleles and their utility. Star alleles are gene haplotypes defined by one or more variants found together in a gene sequence. An official system to assign allele names allows for stable allele identifiers and helps prevent nomenclature duplication in the event that multiple alleles are published at the same time having the same name. Before the mid-1990s, *CYP2D6* alleles were named somewhat arbitrarily in publications with a letter after the gene name, e.g., *CYP2D6A*. However, as the number of discovered *CYP2D6* alleles increased at a rapid rate, it became evident that a better naming system was needed. In 1996, a group of leading gene experts came together to systematically catalog *CYP2D6* alleles and published a nomenclature paper (Daly et al., 1996). Their work was based loosely on a 1987 paper with guidelines for human gene nomenclature and allelic variants that recommended using an asterisk to separate the gene and allele (Shows et al., 1987). A similar system was also used for *NAT2* (Vatsis et al., 1995) and the Human Leukocyte Antigens (HLAs) (Daly et al., 1996). In 2000, Ingelman-Sundberg et al. proposed to extend this standardized nomenclature to cover other cytochrome P450 genes (Ingelman-Sundberg et al., 2000). In 2002, the Human Cytochrome P450 (CYP) Allele Nomenclature Database (CYP DB) was established as a website to catalog and display tables of alleles using the star allele nomenclature (Oscarson & Ingelman-Sundberg, 2002). In 2010, a nomenclature committee for *TPMT* was also established using star allele nomenclature, based on historically named *TPMT* alleles (Appell et al., 2013). PharmVar replaced the CYP DB in 2017 (Gaedigk et al., 2018).

PharmVar now employs a slightly different system for naming alleles that involves a stricter set of naming rules than the original system. The full criteria for designating an allele is available on PharmVar's website at https://www.pharmvar.org/criteria. Haplotypes are grouped into star alleles based on amino acid changes, functional noncoding variants, copy number variants, and structural variants. For a haplotype to receive a new star allele designation, it must contain at least one nucleotide change that results in an amino acid change or have been experimentally shown to affect transcription, splicing, translation, activity, lead to posttranscriptional or posttranslational modifications, or contain a structural rearrangement of the gene. If a haplotype doesn't qualify for a new star allele assignment, it will be assigned as a suballele to an existing star allele.

All suballeles of a star allele contain the defining star allele variants plus one or more additional variants. Suballeles of any given star allele should have the same or similar function due to containing the same functional variants. The extra variants in a suballele should not affect function but they can complicate genotyping assays and have the potential to lead to inaccurate interpretation of results. PharmVar catalogs suballeles to provide the most comprehensible data repository and has developed a standardized numbering system to accommodate potentially large numbers of suballeles for any given star allele. For example, *CYP2D6*2.001* and *CYP2D6*2.002* are different suballeles of *CYP2D6*2* (Fig. 2.2). This comprehensive, publicly available catalog of allelic variation including suballeles is crucial for the design of genetic testing platforms (genotype-based or sequence). It is also valuable to computational algorithms that assign star allele diplotypes from next-generation sequencing (NGS)-based data (e.g., Astrolabe, previously called Constellation (Twist et al., 2017)).

While suballele information can be valuable in the situations discussed above, there is no need to distinguish suballeles when predicting clinical phenotypes from patient diplotypes because all alleles under a star number should have the same function. For this reason, the results of a pharmacogenomic test can be reported as star alleles without the suballele designation (e.g., *CYP2C9*1, *3, *8, etc.*)—also known as "core alleles"—even if a test is capable of distinguishing suballeles. A core allele is defined by variants that are found in all of the suballeles and (1) cause an amino acid change, (2) impact function by changing expression levels, or (3) interfere with splicing. For example, all *CYP2C9*2* suballeles share the c.430C > T (p.R144C)

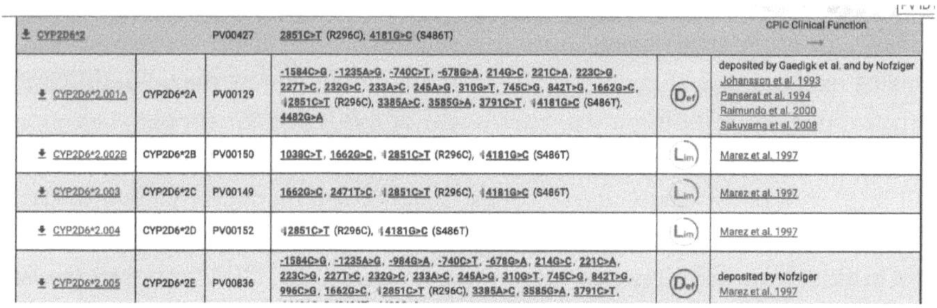

| | | | | | |
|---|---|---|---|---|---|
| CYP2D6*2 | | PV00427 | 2851C>T (R296C), 4181G>C (S486T) | | CPIC Clinical Function |
| CYP2D6*2.001A | CYP2D6*2A | PV00129 | -1584C>G, -1235A>G, -740C>T, -678G>A, 214G>C, 221C>A, 223C>G, 227T>C, 232G>C, 233A>C, 245A>G, 310G>T, 745C>G, 842T>G, 1662G>C, 2851C>T (R296C), 3385A>C, 3585G>A, 3791C>T, 4181G>C (S486T), 4482G>A | Def | deposited by Gaedigk et al. and by Nofziger<br>Johansson et al. 1993<br>Panserat et al. 1994<br>Raimundo et al. 2000<br>Sakuyama et al. 2008 |
| CYP2D6*2.002B | CYP2D6*2B | PV00150 | 1038C>T, 1662G>C, 2851C>T (R296C), 4181G>C (S486T) | Lim | Marez et al. 1997 |
| CYP2D6*2.003 | CYP2D6*2C | PV00149 | 1662G>C, 2471T>C, 2851C>T (R296C), 4181G>C (S486T) | Lim | Marez et al. 1997 |
| CYP2D6*2.004 | CYP2D6*2D | PV00152 | 2851C>T (R296C), 4181G>C (S486T) | Lim | Marez et al. 1997 |
| CYP2D6*2.005 | CYP2D6*2E | PV00836 | -1584C>G, -1235A>G, -984G>A, -740C>T, -678G>A, 214G>C, 221C>A, 223C>G, 227T>C, 232G>C, 233A>C, 245A>G, 310G>T, 745C>G, 842T>G, 996C>G, 1662G>C, 2851C>T (R296C), 3385A>C, 3585G>A, 3791C>T, | Def | deposited by Nofziger<br>Marez et al. 1997 |

**Figure 2.2** *Screenshot of **CYP2D6*2** core allele definition (A) and suballeles (B).* PharmVar catalogs suballeles and has developed a standardized numbering system to categorize suballeles for any given star allele.

variant but no other functional variants in CYP2C9, so this variant is the *CYP2C9*2* core allele definition. However, a sequence variant found in a core allele definition is not necessarily unique to that star allele. For example, c.100C > T (p.34S) is the only variant in the core allele definition for *CYP2D6*10*, but is also found in *CYP2D6*4*, *CYP2D6*6*, and *CYP2D6*36*, among other *CYP2D6* star alleles.

The PharmVar ID is a permanent numeric identifier used in addition to star allele name. Because the functional interpretation of an allele can change over time as more information becomes available, the star allele assignment can potentially change. For example, a suballele can become a separate star allele if it is subsequently determined that a variant functionally changes the allele function from decreased to loss of function. In such cases, PharmVar will revise the allele designation, changing the "star" number, but the PharmVar ID will remain constant. For this reason, PharmVar suggests citing PharmVar IDs in addition to star allele names (e.g., *CYP2D6*46.002* (PV00182)).

PharmVar encourages submissions of alleles from the scientific community. For smaller genes (10–15 kb) such as *CYP2D6*, haplotypes typically can be confirmed experimentally with NGS-based single-molecule sequencing techniques or allele-specific long-range polymerase chain reaction. Sanger sequencing is considered the gold standard but NGS technology has advanced such that PharmVar accepts NGS-based submissions when quality control criteria are fulfilled. However, for larger genes such as *CYP2C9* or *CYP2C19*, haplotype definitions are often based on computationally inferred predictions. PharmVar assigns levels reflecting the strength of evidence for haplotype definitions based on the method used to determine the haplotype. Levels include "definitive" (e.g., Sanger sequencing) and "moderate" (e.g., computational inference). Before the creation of PharmVar's criteria, many star allele definitions lacked crucial pieces of information that would now be required. As a result, these haplotypes are considered to have "limited" evidence. PharmVar accepts submissions of existing star alleles assigned an evidence level less than "definitive" with the goal of building a stronger body of evidence for them.

The PharmVar website displays star allele definitions by gene and allelic data displayed consistently across genes. Variants are mapped by gene to current versions of gene and transcript reference sequences (RefSeq) and multiple builds from the Genome Reference Consortium, allowing users to use both individual gene transcripts and entire genomic sequences as references. The variant view available on PharmVar displays coordinates across all references using two different numbering systems: (1) counting from the

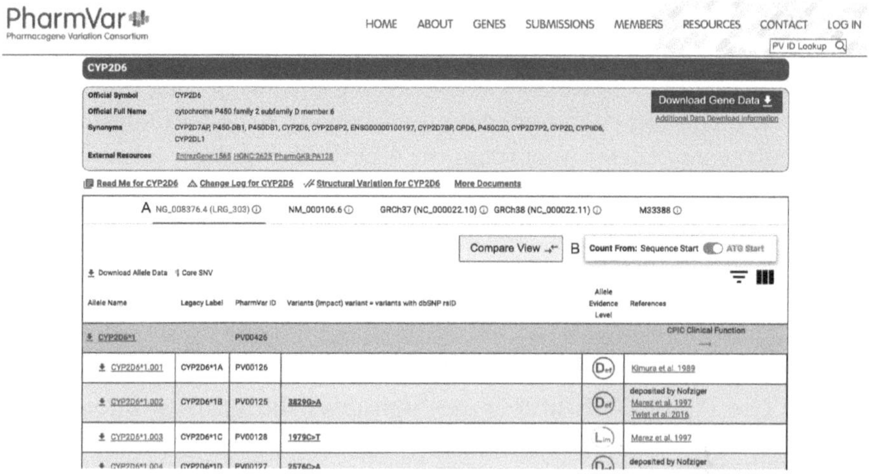


**Figure 2.3** The PharmVar webpage for CYP2D6. Variant positions can be viewed based on the reference sequences listed at the top of the table (e.g., NG_008376.4) (A) and numbering can be flipped between Sequence Start and ATG Start by toggling the slider (B). *Screenshot taken at https://www.pharmvar.org/gene/CYP2D6 on September 1st, 2021.*

sequence start, and (2) counting from the ATG start codon (Fig. 2.3). The use of multiple reference sequences and numbering methods makes the resource user friendly, as a wide number of alternate terms for a desired variant's name can be used to access information for that variant. The literature and testing labs do not consistently refer to variants using the same reference sequence or numbering and PharmVar allows for convenient mapping across all these possibilities. The variant page lists all haplotypes containing the variant and provides the rsID with a link to the dbSNP database. The user can query all alleles carrying a particular variant of interest. PharmVar displays functional consequences of allelic variation in the gene tables, which are synchronized with CPIC assigned allele function that are also found in the PharmGKB gene information tables. In addition to the interactive tabular display for each gene, PharmVar provides comprehensive and explanatory supporting "Read Me" documents and a log of all changes and edits for each gene.

## Pharmacogenomics knowledgebase

The Pharmacogenomics Knowledgebase (PharmGKB; https://www.pharmgkb.org) is the premiere freely available centralized online pharmacogenomics resource (Whirl-Carrillo et al., 2012, 2021). Its mission is to curate, annotate, integrate, and disseminate pharmacogenomic knowledge. PharmGKB curates this knowledge from a variety of sources, including peer-reviewed publications, drug labels from multiple regulatory agencies (including the FDA), and genotype-based drug dosing guidelines from expert societies. PharmGKB serves diverse user groups from students to professionals in the clinical and scientific communities, and provides comprehensive and integrated drug, gene, variant, and disease information to pharmacologists, clinical investigators, biologists, and informaticians as well as to the general public. All PharmGKB data are freely available through a Creative Commons license. The PharmGKB homepage highlights the primary interests of most users for easy browsing and also contains a "Google-like" search bar for specific queries. PharmGKB provides resources for those users new to the pharmacogenomics field as well as those already knowledgeable about pharmacogenomics but new to PharmGKB.

PharmGKB utilizes both manual and automatic curation: PharmGKB expert scientists manually curate peer-reviewed publications, and these publications are also automatically identified through use of PGxMine, a supervised machine learning pipeline that extracts variant—drug associations from PubMed abstracts and full text articles in PubMed Central (Lever et al., 2020). PharmGKB's routine manual curation includes a varied set of high-profile journals as well as pharmacogenomics-focused journals. Additional publications are also manually curated as part of pathway and Very Important Pharmacogene (VIP) summary production (described below) or other specific projects. Publications identified though the PGxMine tool also feed into the manual curation process. Publications sent to PharmGKB by website users are also manually curated when relevant.

To manually curate peer-reviewed publications, PharmGKB scientists read them and select specific pieces of information to populate defined fields in the PharmGKB database, such as study size and statistics (e.g., *P*-value, odds ratio). The scientists also annotate these articles using standardized terminologies (Table 2.1) and formats so that the PGx information is machine-readable and can be deposited into the database in a structured manner, allowing for indexing, searching, and retrieval by users. PharmGKB has created web-based tools to aid and provide structure for PharmGKB scientists in the curation and annotation process.

**Table 2.1** Standardized vocabularies used by PharmGKB.

| Information type | Standardized terminology |
|---|---|
| Gene name | HUGO Gene Nomenclature Committee (HGNC) |
| Variant | dbSNP rs identifier (rsID) |
| Cytochrome P450 variant allele | PharmVar core star (*) alleles |
| Drug | PubChem, RxNorm |
| Disease/phenotype | Medical Subject Heading (MeSH), SNOMED CT |
| Allele function | CPIC allele function terms |
| Gene phenotype (e.g., metabolizer status) | CPIC gene phenotype term |
| Publication | PubMed identifier (PMID), PubMed central identifier (PMCID) |

## PharmGKB annotations

There are two types of basic PharmGKB annotations of published literature: literature annotations and variant annotations. In both cases, the annotation includes a PubMed identifier (PMID), allowing users to connect the annotation to the original literature source.

A *literature annotation* captures the genes and drugs discussed in a publication using standardized terminologies. It is the simplest type of annotation on PharmGKB and only indicates co-occurrence of the terms in a publication. A literature annotation may also capture if there is a direct relationship between genes and drugs as reported in the publication and of what "phenotype category" the relationship is a type: PD, dosage, efficacy, toxicity, metabolism/PK, or other.

A *variant annotation* captures more information from a publication than a literature annotation, including the specific variant or allele that is associated with a drug response phenotype and in what gene (if any) the variant is found or associated with (variants may be upstream or downstream of the gene boundaries but affect gene function). Variant names are standardized using the dbSNP rs identifier (rsID) when available, otherwise the genomic position on a designated reference sequence is used. For genes where a star (*) allele nomenclature is commonly used to denote gene haplotypes or variants, PharmGKB uses an authoritative source to standardize the allele terminology when available. For cytochrome P450s and some other pharmacogenomic genes, that source is PharmVar. Other sources include the TMPT Nomenclature Committee (https://liu.se/en/research/tpmt-nomenclature-committee)

and the UGT1A1 Nomenclature (https://www.pharmacogenomics.pha.ulaval.ca/wp-content/uploads/2015/04/UGT1A1-allele-nomenclature.html). Diseases, adverse events, and phenotypes use standard vocabularies, such as MeSH and SNOMED CT terms. A sentence is constructed using one of three formulas available to PharmGKB scientists in the web-based curation tool. The three types of sentences represent three different kinds of PGx results: (1) variant—drug—phenotype, (2) variant—drug, and (3) functional assay. The standardized sentence begins with a variant or genotype, indicates whether or not an association was found (PharmGKB also annotates publications which report negative results) and the direction of the association (e.g., "increased" dosage). The sentence can include the population in the study (e.g., "in people with diabetes"). The author-reported significance of the association is also recorded. The curator has the option of using free text to report any other information that would be important context for the reader. A variant annotation can also provide information about the study type (e.g., GWAS, case-control), study size, statistics reported (e.g., association *P*-value, odds ratio), and study population ethnicity. To capture study population ethnicity, PharmGKB uses a system of nine biogeographical groups developed in collaboration with expert population geneticists (Huddart, Fohner, et al., 2019a).

All annotated and captured information from a publication in a variant annotation is presented for users in tabular form both on the website display and as downloadable tab separated value (TSV) files that can be filtered and searched. The goal of variant annotations is to summarize and present the relevant information about a particular variant—drug association from each publication to the user. In some cases, there are many publications which report on the same variant—drug combination. In many cases, these publications agree and build a body of evidence supporting the association. However, sometimes some studies are unable to replicate an association or report an association in a different direction than others (e.g., "increased metabolism" vs. "decreased metabolism"). In situations with many variant annotations, it can be onerous for the user to review all and determine where all the evidence points. PharmGKB also provides curated summaries of variant—drug associations, called clinical annotations, which take the burden of aggregating this knowledge off of the user.

*Clinical annotations* summarize the information from variant annotations for a particular variant—drug—phenotype. A "Level of Evidence" (LOE) is assigned to each clinical annotation to indicate the amount of literature support annotated for the variant—drug association.

Clinical annotations are one of the most popular features on PharmGKB and have used to annotate the human genome (Ashley et al., 2010; Chen et al., 2012; Dewey et al., 2011, 2014), develop PGx testing panels (Johnson et al., 2012), and report PGx testing results to clinicians and patients.

Genetic variation with a drug phenotype can be reported for a variant located at a particular genomic position (e.g., single nucleotide polymorphism, indel) or for a star allele, which represents variation across the gene described above. Accordingly, there are two types of clinical annotation: variant-level and gene-level. Variant-level annotations provide genotype-based summaries for a specific rsID. Usually, variant options are binary: a person either has the variant or the reference sequence at that position. There are, however, examples of tri-allelic variation or multiple insertion/deletion/repeat variation at a genomic position. A clinical annotation reflects all variant possibilities that have been curated from the literature. Likewise, gene-level annotations provide allele-based summaries for a gene. In this case, a clinical annotation reflects all the star alleles curated from the literature for that gene. Both variant-based and gene-based summaries are written based on all variant annotations for that variant—drug—phenotype or gene—drug—phenotype, including those that support as well as those that contradict the association. Variant annotations are displayed on the PharmGKB website under corresponding clinical annotations for transparency so that the user can easily access information on the individual studies the clinical annotation is based on.

Annotated pharmacogenomic guidelines or drug labels (see below for more information about annotation types) can be added as additional support to clinical annotations if they contain information about the relevant variant—drug—phenotype association. These supporting pieces of evidence will be displayed below the clinical annotation on the website along with the supporting variant annotations.

## PharmGKB levels of evidence

PharmGKB uses a quantitative LOE system to signal to users the amount of supporting scientific evidence for a clinical annotation (Fig. 2.4) (Whirl-Carrillo et al., 2021). Levels include high (Level 1), moderate (Level 2), low (Level 3), and associations unsupported by the evidence (Level 4). The description and criteria for each level is found at https://www.pharmgkb.org/page/clinAnnLevels. The LOE is automatically set based on the clinical annotation's score, which is calculated based on the

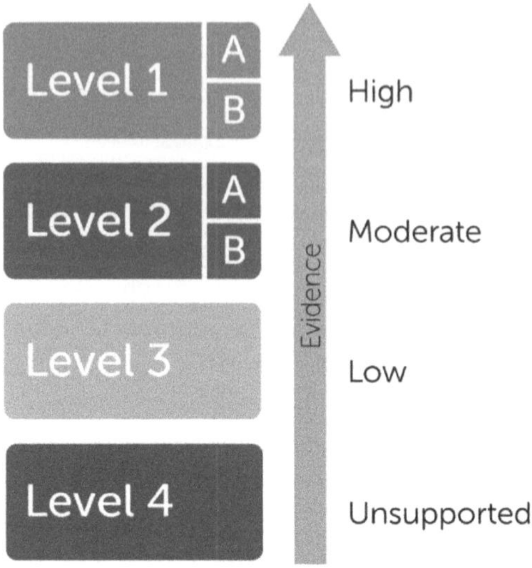


**Figure 2.4** PharmGKB level of evidence categories. PharmGKB Level of Evidence categorizes Clinical Annotations based on the amount of annotated evidence supporting an association between a variant and drug phenotype, ranging from High to Unsupported (i.e., the evidence does not support an association).

supporting variant, pharmacogenomic guideline, and drug label annotations. First, variant annotations are assigned a score based on specific attributes (see https://www.pharmgkb.org/page/varAnnScoring for details) and the scores for all variant annotations included in a clinical annotation are summed together (Fig. 2.5). Variant annotations that report associations in an

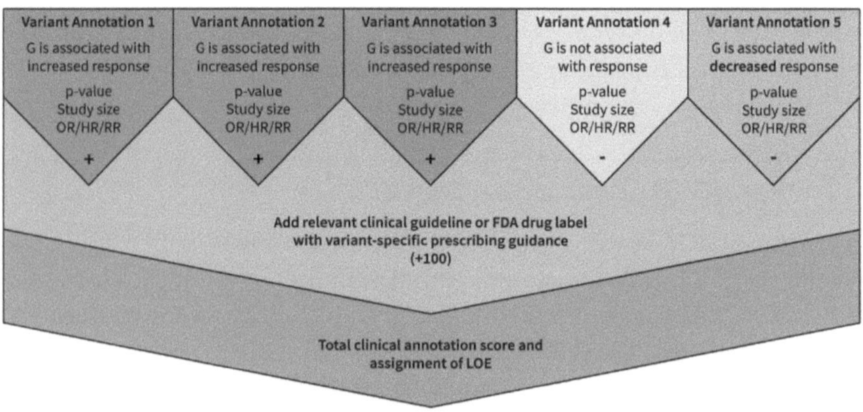


**Figure 2.5** PharmGKB level of evidence calculation. The Level of Evidence is based on the score that is calculated for all annotated evidence regarding the association, including Variant Annotations and any annotated FDA drug labels or PGx guidelines.

opposite direction to other variant annotations in the clinical annotation are considered "conflicting" (e.g., a variant annotation that associates a variant with increased response to a particular drug, while other variant annotations associate the same variant with decreased response to the same drug). Scores from any conflicting variant annotations are subtracted from the clinical annotation score during the summation process. Then, relevant drug label annotations and pharmacogenomic guideline annotations contribute to the clinical annotation score with a flat increase. The clinical annotation score is only used to determine the LOE and not intended to be used to rank or compare clinical annotations within a LOE category.

In some cases, certain gene alleles may be less well represented in the evidence base than others, typically because they are not frequently tested. To address this, PharmGKB uses a "Limited Evidence" tag that denotes individual alleles on gene-level annotations which are supported by notably less evidence than other alleles. Use of the tag does not alter the assigned LOE of the clinical annotation. Level 1 and 2 clinical annotations are independently reviewed by a second curator. Changes to clinical annotations are logged in the history section which is displayed to users at the bottom of each annotation page.

Clinical annotations are examples of integrated pharmacogenomic information based on curated and annotated primary literature, pharmacogenomic guidelines, and drug labels. Scientific evidence from multiple sources is combined and interpreted by PharmGKB scientists to create new knowledge. Other examples of integrated scientific evidence at PharmGKB are the drug-centric PK and PD pathways and Very Important Pharmacogene (VIP) summaries. Both pathways and VIPs are created from the integration of curated primary literature discovered from extensive literature reviews of the drugs or genes.

## PharmGKB pathways

*PharmGKB pathways* are evidence-based illustrations of the PK and/or PD of a drug with relevant or potential pharmacogenomic associations. Drugs are selected for PharmGKB pathways through extensive review of a variety of sources, including annotated regulatory agency–approved drug labels and PGx guidelines. Pathways diagrams are accompanied by a written overview with details of the PK or PD process and other important pharmacogenomic-related information. Interactions within each pathway are supported by manually curated evidence from published literature.

The overarching pathway is essentially pieced together from information from multiple publications, and the specific references providing support for each arrow in the pathway are made available on the "Components" tab of the pathway webpage. Information contained within each pathway is available for download in TSV, BioPAX (Demir et al., 2010) and Graphical Pathway Markup Language (GPML) formats. The pathway graphic is available for download as a Portable Document Format (PDF) file.

## Very important pharmacogene summaries

*Very Important Pharmacogene (VIP)* summaries give an overview of a significant gene involved in metabolism of, or response to, one or more drugs. Often, VIPs either play a role in the metabolism of many drugs (e.g., *CYP2D6*) or contain variants which contribute to a severe drug response (e.g., *HLA-B*). VIP summaries usually include background information on the gene, any disease associations and in-depth information about the gene's pharmacogenomics. PharmGKB began creating VIPs in collaboration with members of the Pharmacogenomics Research Network (PGRN) in 2006 and the original VIP list was based on PGRN gene nominations. The list grew based on several sources, including PharmGKB clinical annotations and annotated regulatory agency approved drug labels and pharmacogenomic guidelines, and many VIP summaries have been published in the journal *Pharmacogenetics and Genomics*. In 2020, an evidence assessment was undertaken by PharmGKB and the list of VIPs was divided into three tiers. *Tier 1* contains genes with substantial evidence to support their importance in pharmacogenomics, including prescribing guidance from annotated pharmacogenomic guidelines. These Tier 1 VIPs are used in the assignment of Level 2A to clinical annotations. *Tier 2* contains genes previously nominated as VIPs that have more limited evidence to support their importance in pharmacogenomics. VIP genes that are important in tumor pharmacogenomics are assigned to the Cancer Genome group.

## Drug label annotations

PharmGKB also manually curates and annotates (1) drug labels from multiple regulatory agencies (see below) and (2) genotype-based drug dosing guidelines from expert societies using the same manual curation process used for peer-reviewed publications. PharmGKB captures specific pieces of information in a structured manner using web-based tools and annotates the evidence with standardized terminologies (Table 2.1), allowing for database storage and retrieval for website presentations and downloadable files.

PharmGKB curates FDA-approved drug labels based on the information posted on the FDA's Table of Pharmacogenomic Biomarkers in Drug Labeling and the Table of Pharmacogenetic Associations. PharmGKB annotates FDA-approved labels with pharmacogenomic information not listed in these tables as these labels are brought to the attention of curators (Fig. 2.6). PharmGKB annotations of drug labels have expanded and to include labels containing pharmacogenomic information from the European Medicines Agency and Health Canada as these labels brought to curators' attention. PharmGKB has also partnered with collaborators in one-off projects to curate labels from Swissmedic (Jeiziner et al., 2021) and Japan's Pharmaceuticals and Medical Devices Agency. PharmGKB annotations provide a brief summary of the pharmacogenomic data in the label, an excerpt from the label, and a downloadable highlighted label PDF file. Labels are annotated with drugs, genes, and variants (if applicable) using the standardized terminologies in Table 2.1. Labels from multiple regulatory drug agencies have been annotated; users can access the full annotation for any given label by clicking on the colored tag in the grid based on drug by agency. This landing table shows the PharmGKB-assigned PGx level and other tags indicating what PGx information is available before clicking on to the full annotation.

The amount and type of pharmacogenomic information on the labels from the two FDA tables span a wide range, from black box warnings to mentions of metabolism in the *Pharmacokinetics* section of the label. As a few labels actually state that variation in genes/enzymes is not clinically

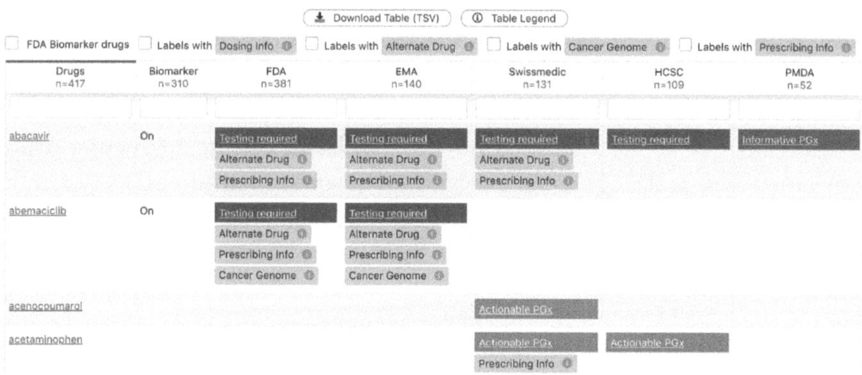


**Figure 2.6** Screenshot of PharmGKB landing page listing all annotated drug labels. This image shows the PharmGKB landing page listing all annotated drug labels.

relevant and other labels for selected drugs on the Table of Pharmacogenetic Associations do not mention a gene, PharmGKB has created a "Pharmacogenomics Level" system to indicate which labels require or recommend pharmacogenomic testing and which have information that could be clinically relevant, for users' reference. The levels include "Testing Required," "Testing Recommended," "Actionable PGx," and "Informative PGx." Full definitions for each category are found on the PharmGKB website (https://www.pharmgkb.org/page/drugLabelLegend#pgx-level). Users should note that the Pharmacogenomics Level is the PharmGKB interpretation of the level of action implied in each label and is not endorsed by the FDA.

In recent years, some FDA-approved drug labels have begun to include more specific pharmacogenomic information. The *Prescribing* section of the PharmGKB annotation captures guidance from the label for patients with a particular genotype/metabolizer phenotype, if available. For example, the FDA-approved dronabinol label states, "Monitoring for increased adverse reactions is recommended in patients known to carry genetic variants associated with diminished CYP2C9 function." The label is also annotated with tags to indicate if it provides dosing information (e.g., "In patients who are poor CYP2D6 metabolizers, the total daily dosage of [deutetrabenazine] should not exceed 36 mg") or states that a drug is either indicated or contraindicated based on genotype or metabolizer phenotype. Information from the Table of Pharmacogenetic Associations is included on the drug label annotation when applicable.

Many drugs on the FDA's Table of Pharmacogenomic Biomarkers in Drug Labeling are cancer therapies indicated or contraindicated for specific biomarkers found in tumor or other cancer cells rather than germline variation. This is a particular subset of pharmacogenomics that requires special testing rather than the more typical genetic analysis. PharmGKB annotates these labels as "Cancer Genome" for quick and easy identification.

Individual drug label annotations are displayed on the respective drug and gene pages, and a comprehensive list of all annotated labels are available from the PharmGKB homepage and as a downloadable tab TSV file that can be filtered and searched.

## Pharmacogenomic guideline annotations

PharmGKB curates and annotates genotype-based drug dosing guidelines published by CPIC, DPWG, and other professional societies.

PharmGKB pharmacogenomic guideline annotations present a brief summary of the genotype-based dosing recommendations from these groups annotated with the respective drugs, genes, and variants using standardized terminology from Table 2.1. They highlight the prescribing recommendations based on a patient's genotype or metabolizer phenotype, and include excerpts from the guideline and links to the underlying evidence source. PharmGKB has also created video explanations of a number of CPIC guidelines, embedded from a dedicated YouTube channel. For CPIC and some DPWG guidelines, PharmGKB provides functionality for users to "look-up" recommendations for specific genotypes. Interactive pull-down menus allow users to enter genotypes in order to display the corresponding inferred phenotype (e.g., ultra-metabolizer), associated prescribing recommendation, and strength of the recommendation. PharmGKB provides this functionality via online tools created to allow PharmGKB scientists to manually enter the information provided by CPIC and DPWG in supplemental tables containing allele function, diplotype-to-phenotype mapping, and phenotype-to-recommendation mapping into the database. Underlying code computationally traces from alleles in the entered genotype to recommendations (Fig. 2.7).

Regardless of annotation type, all annotations which contain information from or are particularly relevant to the pediatric community are tagged on the PharmGKB website with a pediatric icon. For those users with a particular interest in that specialty population, PharmGKB has a pediatric specific available portal at https://pediatric.pharmgkb.org. All annotation types can be viewed on the PharmGKB website, queried through the PharmGKB API, or downloaded in TSV formats (https://www.pharmgkb.org/downloads). In addition to the various annotation types, PharmGKB provides other resources, including PGx publication tips, a cancer pharmacogenetics webpage, and translational information from allele nomenclature to predicted phenotypes created in collaboration with PharmVar and CPIC.

## Dutch Pharmacogenetics Working Group of the Royal Dutch Pharmacists Association (KNMP)

One of the pharmacogenomic guideline sources curated by PharmGKB is the Dutch Pharmacogenetics Working Group (DPWG) of the Royal Dutch Pharmacists Association. The DPWG was established in 2005 by the Royal Dutch Pharmacist's Association (KNMP) as a multidisciplinary group including clinical pharmacists, physicians, clinical pharmacologists, clinical

## Annotation of CPIC Guideline for amitriptyline and CYP2C19, CYP2D6

### Summary

The CPIC Dosing Guideline update for amitriptyline recommends an alternative drug for CYP2D6 ultrarapid or poor metabolizers and CYP2C19 ultrarapid, rapid or poor metabolizers. If amitriptyline is warranted, consider a 50% dose reduction in CYP2D6 or CYP2C19 poor metabolizers. For CYP2D6 intermediate metabolizers, a 25% dose reduction should be considered.

### Specify a genotype for specific annotations

**Pick a CYP2C19 Genotype**

-- -- 

**Pick a CYP2D6 Genotype**

-- -- 

### Annotation

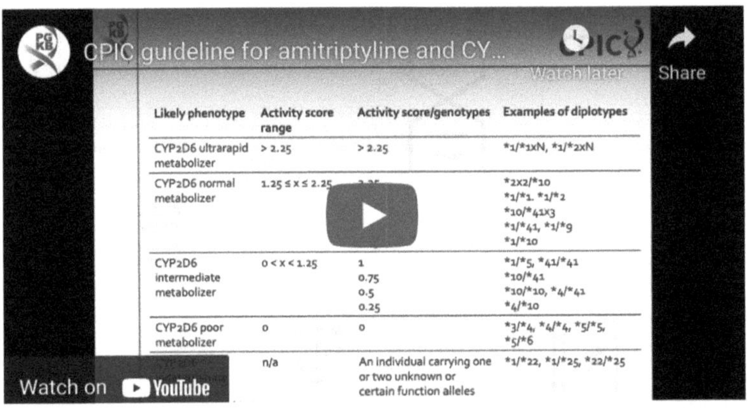


**Figure 2.7** PharmGKB annotation for the cpic guideline regarding amitriptyline and ***CYP2C19*** and ***CYP2D6***. When the user enters genotypes for CYP2D6 and CYP2C19 in the pull-down menus, the corresponding CPIC prescribing recommendations are displayed. A short video summarizes the highlights of the guideline and additional information is provided.

chemists, epidemiologists, and toxicologists. Their objectives are to develop pharmacogenetics-based therapeutic (dose) recommendations and to assist drug prescribers and pharmacists by integrating the recommendations into computerized systems for drug prescription and automated medication surveillance.

The DPWG published two papers with pharmacogenomic therapeutic recommendations in 2008 and 2011 (Swen et al., 2008, 2011). PharmGKB collaborated with DPWG in 2013 to post recommendations from these publications. DPWG methods included doing a systematic search for each drug. The articles included in the reference lists were individually screened for additional material or papers. Wherever information relating to gene–drug interaction was present in the European Public Assessment Report, the manufacturer was asked to provide further details. Review articles, studies involving non-human subjects, and in vitro experiments were excluded. Two parameters were defined for each article retrieved: (1) LOE of the gene–drug interaction and (2) clinical relevance of the potential adverse drug event, decreased therapeutic response, or other clinical effect resulting from the gene–drug interaction. The LOE of the gene–drug interaction indicated the quality of the evidence found in literature for the gene–drug interaction. The LOE was scored on a five-point scale with a range from 0 (lowest evidence) to 4 (highest evidence). The clinical relevance was scored on a seven-point scale derived from the National Cancer Institute's Common Toxicity Criteria. A clinical or PK effect that was not statistically significant was classified as AA (lowest impact), whereas death, for example, was classified as F (highest impact).

DPWG recommendations are now posted as a PDF file on the Royal Dutch Pharmacists Association (KNMP) website at https://www.knmp.nl/patientenzorg/medicatiebewaking/farmacogenetica/pharmacogenetics-1/pharmacogenetics. Recommendations are standard in the Netherlands and made available to others via the download file. The file is updated after each meeting of the DPWG, approximately every 3 months, but not all drug recommendations change with any given release. DPWG also makes available additional information about genes in their recommendations such as allele activity and translation of genotype to phenotype groups. The Ubiquitous Pharmacogenomics (U-PGx) project aims to enable widespread adoption of the DPWG guidelines across Europe (https://upgx.eu/guidelines/).

## Clinical Pharmacogenetics Implementation Consortium

The Clinical Pharmacogenetics Implementation Consortium (CPIC; https://cpicpgx.org) is a global consortium focused on providing prescribing guidance to practicing clinicians based on patient genotypes

(Relling et al., 2019; Relling & Klein, 2011). Genotype-based drug dosing guidelines are created by clinical and pharmacogenomic experts following extensive literature reviews and evidence grading in accordance with transparent standard operating procedures. In support of these guidelines, CPIC provides translational information from allele nomenclature to predicted phenotypes (created in collaboration with PharmVar and PharmGKB), example clinical decision support (CDS) language, and possible electronic health record (EHR) workflows. Data from the guideline manuscripts and supporting materials are available in structured formats via the CPIC database and the API. CPIC guidelines operate on the key assumption that clinical preemptive genotyping will become more widespread, and clinicians may be presented with patients' genotypes even if they have not explicitly ordered a test for a specific drug. Therefore, CPIC guidelines are designed to help clinicians understand how available genetic test results should be used to optimize drug therapy, instead of whether or when tests should be ordered. CPIC's guidelines have been endorsed by the American Society for Clinical Pharmacology and Therapeutics (ASCPT) and the American Society of Health Systems Pharmacists (ASHP).

CPIC guidelines are based on extensive literature reviews (including PubMed searches, manual curation, and annotation of the search results) and review and grading of curated information and publications by guideline authors. Literature evidence can include in vitro enzyme activity of protein variants, in vitro enzyme activity from tissues isolated from individuals of known genotypes, in vivo PK and PD data, and in vivo preclinical and clinical PK and PD studies. The CPIC review process applies a standard system for grading levels of evidence linking genotypes to phenotypes (https://cpicpgx.org/levels-of-evidence/). Guideline authors determine how to assign clinical phenotypes to genotypes (represented as a mapping table in the guideline) and create prescribing recommendations typically based on the clinical phenotype (Figs. 2.8 and 2.9).

For some genes, such as *CYP2D6*, the phenotype categories are insufficiently specific for genotype-based recommendations and the diplotype activity score (Gaedigk et al., 2007) is used in these cases. CPIC uses a standard system to assign a level of strength to each prescribing recommendation. This system is largely adopted from the transparent and simple rating scale for evidence-based recommendations regarding the use of antiretroviral agents (http://aidsinfo.nih.gov/contentfiles/AdultandAdolescentGL.pdf) with slight modification, and includes strong, moderate, and optional categories. A fourth category of "no recommendation" was added to account for

| Likely phenotype[a,b] | Activity score | Genotypes | Examples of diplotypes |
|---|---|---|---|
| Normal metabolizer | 2 | An individual carrying two normal function alleles | *1/*1 |
| Intermediate metabolizer | 1.5<br>1 | An individual carrying one normal function allele plus one decreased function allele; OR<br>one normal function allele plus one no function allele OR two decreased function alleles | *1/*2<br>*1/*3, *2/*2 |
| Poor metabolizer | 0.5<br>0 | An individual carrying one no function allele plus one decreased function allele; OR<br>two no function alleles | *2/*3<br>*3/*3 |
| Indeterminate | n/a | An individual carrying allele combinations with uncertain and/or unknown function alleles | *1/*7, *1/*10, *7/*10, *1/*57 |

[a]Assignment of allele function and associated citations can be found at https://www.pharmgkb.org/page/cyp2c9RefMaterials (see *CYP2C9* Allele Definition Table and *CYP2C9* Allele Functionality Table in refs. 1,2). For a complete list of *CYP2C9* diplotypes and resulting phenotypes, see the *CYP2C9* Genotype to Phenotype Table in refs. 1,2. [b]See the *CYP2C9* Frequency Table in refs. 1,2 for population-specific allele and phenotype frequencies.

**Figure 2.8** Assignment of likely CYP2C9 phenotypes based on genotype. Prescribing recommendations are typically based on the clinical phenotype. *From Theken, K. N., Lee, C. R., Gong, L., Caudle, K. E., Formea, C. M., Gaedigk, A., Klein, T. E., Agúndez, J. A. G. & Grosser, T. (2020). Clinical Pharmacogenetics implementation consortium guideline (CPIC) for CYP2C9 and nonsteroidal anti-inflammatory drugs.* Clinical Pharmacology and Therapeutics, 108*(2), 191–200. https://doi.org/10.1002/cpt.1830*

| Phenotype[a] | Implication | Therapeutic recommendation | Classification of recommendation | Other considerations |
|---|---|---|---|---|
| CYP2C9 normal metabolizer | Normal metabolism | Initiate therapy with recommended starting dose.<br>In accordance with the prescribing information, use the lowest effective dosage for shortest duration consistent with individual patient treatment goals. | Strong | |
| CYP2C9 intermediate metabolizer AS of 1.5 | Mildly reduced metabolism | Initiate therapy with recommended starting dose.<br>In accordance with the prescribing information, use the lowest effective dosage for shortest duration consistent with individual patient treatment goals. | Moderate | IMs might have a higher than normal risk of adverse events especially in individuals with other factors affecting clearance of these drugs, such as hepatic impairment or advanced age. Further caution should be taken with ibuprofen use in individuals carrying the *CYP2C9*2* allele as it is in linkage disequilibrium with *CYP2C8*3* and ibuprofen is also metabolized by CYP2C8. |
| CYP2C9 intermediate metabolizer AS of 1 | Moderately reduced metabolism; higher plasma concentrations may increase probability of toxicities | Initiate therapy with lowest recommended starting dose. Titrate dose upward to clinical effect or maximum recommended dose with caution.<br>In accordance with the prescribing information, use the lowest effective dosage for shortest duration consistent with individual patient treatment goals. Carefully monitor adverse events, such as blood pressure and kidney function during course of therapy | Moderate | IMs might have a higher than normal risk of adverse events especially in individuals with other factors affecting clearance of these drugs, such as hepatic impairment or advanced age. Further caution should be taken with ibuprofen use in |

**Figure 2.9** Therapeutic recommendations for selected non-steroidal anti-inflammatory drugs based on CYP2C9 phenotype. Partial Table 2 from the guideline providing therapeutic recommendation and strength by *CYP2C9* phenotype. *Excerpted from Theken, K. N., Lee, C. R., Gong, L., Caudle, K. E., Formea, C. M., Gaedigk, A., Klein, T. E., Agúndez, J. A. G. & Grosser, T. (2020). Clinical pharmacogenetics implementation consortium guideline (CPIC) for CYP2C9 and nonsteroidal anti-inflammatory drugs.* Clinical Pharmacology and Therapeutics, 108*(2), 191–200. https://doi.org/10.1002/cpt.1830*

phenotypes where the evidence is insufficient for CPIC to provide a recommendation to guide clinical practice. Definitions for the rating system are found at https://cpicpgx.org/strength-of-recommendations/. Each CPIC guideline manuscript adheres to a standard format and the process for guideline creation has been published (Caudle et al., 2014).

CPIC provides supporting materials accompanying each guideline which can aid in clinical implementation, from pharmacogenomic test results to electronic health record (EHR) integration. The types of files and their contents are described below.

## Allele definition table

For genes where star (*) allele nomenclature is commonly used to denote gene haplotypes or variants, CPIC coordinates with authoritative sources to define the allelic variants. For cytochrome P450s and some other pharmacogenomic genes, CPIC allele definition tables are based on PharmVar core alleles with the addition of overlapping variants that may be found in suballeles. Other sources used by CPIC include the TMPT Nomenclature Committee (https://liu.se/en/research/tpmt-nomenclature-committee) and the UGT1A1 Nomenclature (https://www.pharmacogenomics.pha.ulaval.ca/wp-content/uploads/2015/04/UGT1A1-allele-nomenclature.html). The allele definition table contains mappings of variants to positions on multiple reference sequences, DNA base changes, amino acid changes, and rsIDs, when applicable.

## Allele functionality table

CPIC assigns function terms to alleles based on literature evidence reviewed by a panel of researchers, clinicians, pharmacists, and other experts in pharmacogenomics. CPIC puts an emphasis on clinical action when determining allele function, with the goal that allele function assignment should lead to a phenotype assignment that can drive clinical prescribing actionability. An allele's CPIC-assigned function is an "allele clinical functional status" and can differ from the biochemical function reported in the literature. In some cases, a biochemical function can be increased or decreased under experimental conditions but not substantially sufficiently to affect clinical phenotype or actionability, and as a result, may be classified by CPIC as a normal function allele. The standardized terminology for allele function terms used by CPIC have been published and include increased

| GENE: CYP2C9 | | | | | | | |
|---|---|---|---|---|---|---|---|
| Allele/cDNA/rsID | Activity Value (Optional) | Allele Biochemical Functional Status (Optional) | Allele Clinical Functional Status (Required) | Allele Clinical Function Substrate Specificity (Optional) | PMID (Required) | Strength of Evidence (Required) | Summary of Findings (Required) |
| *1 | 1.0 | Normal function | Normal function | | | | |
| *2 | 0.5 | Decreased function | Decreased function | | 29283396; 20150829; 1563752; 12742136; 10413320; 27179628; 25775139; 19082874; 30088221; 15637526 | Definitive | CYP2C9*2 is assigned decreased function based on definitive evidence in hom heterozygous patients and in vitro experimental data. CYP2C9*2 has been stuc than 20 years and its association with impaired function compared to wildtype is established (29283396, 20150829, 15637526). The phenotype for patients hoı the CYP2C9*2 variant was updated from poor metabolizer to intermediate metal reevaluation of prior and new evidence demonstrating CYP2C9*2 has more enz than CYP2C9*3, which results in in a similar phenotype for CYP2C9*1/*3 and C patients (12742136, 10413320, 27179628, 25775139, 19082874, 30088221), can be distinguished from those homozygous for CYP2C9*3/*3. Therefore, cons experts was decreased function with an activity value of 0.5 based on definitive |
| *3 | 0 | Decreased function | No function | | 8809086; 29283396; 20150829; 15637526; 12742136; 10413320; 27179628; 25775139; 19082874; 30088221 | Definitive | CYP2C9*3 is assigned no function based on definitive evidence in homozygous heterozygous patients and in vitro experimental data. CYP2C9*3 has been stuc than 20 years and its association with impaired function compared to wildtype is established (8809086, 29283396, 20150829, 15637526). This allele was previ "decreased function" but after reevaluation of prior and new evidence, CYP2C9 "no function" because its activity is much more decreased compared to CYP2C9 substrates, resulting in the lowest possible designation for the phenotypes obse homozygous patients (12742136, 10413320, 27179628, 25775139, 19082874 Therefore, consensus among experts was no function with an activity value of 0 definitive evidence. |
| | | | | | | | CYP2C9*4 is assigned decreased function based on limited evidence in heteroz |

**Figure 2.10** Sample CPIC allele functionality table for CYP2C9. For each star allele, a clinical function, biochemical function, and activity value is assigned. The literature upon which the function assignments are based are given, along with a summary and the strength of evidence.

function, normal function, decreased function, no function, uncertain function, and unknown function (Caudle et al., 2017).

In addition to the function term, the allele functionality file contains a summary of the literature evidence reviewed per allele, including corresponding PMIDs, as well as the strength of evidence for the allele function assignment, described using a tiered system ranging from no evidence to definitive evidence (Fig. 2.10). The tiered system sets the threshold for the minimum evidence required to assign a clinical function. The strength reflects the type and amount of evidence supporting the function assignment (e.g., results from well-controlled clinical studies, metaanalyses, case reports, preclinical in vivo studies, preclinical in vitro studies, or based on computational predictions of function). Alleles for which the evidence is not sufficient to support function assignment are categorized as "uncertain function." Alleles for which no evidence regarding their functional status can be found during the literature review are categorized as "unknown function." For genes such as CYP2D6 where the clinical phenotype corresponds with an "activity score" (discussed further in the diplotype-phenotype table section below), CPIC uses an "activity value" (AV) to describe a graded activity for alleles. The CPIC standard process for allele function assignment is made available on the "Resources" tab of the CPIC website. Allele functionality tables are reviewed and updated as necessary each time guidelines are published or updated for the gene.

## Diplotype–phenotype table

CPIC provides a table that maps every possible combination of alleles (diplotype) in the allele functionality table for a gene to a phenotype term.

The allele functionality table and the mapping table created by the guideline authors discussed above are used together to generate phenotype terms for every diplotype. Phenotype terms for major drug-metabolizing enzymes and some additional pharmacogenes such as HLA genes were defined in the first CPIC Term Standardization Project (Caudle et al., 2017). Metabolizing enzyme phenotype terms include poor metabolizer, intermediate metabolizer, normal metabolizer, rapid metabolizer, and ultrarapid metabolizer, though not all genes will necessarily use all phenotype terms (e.g., there are no rapid or ultrarapid *CYP2C9* metabolizers). The assignment of a phenotype mitigates the complexity of interpreting genetic test results into clinical actionability. For some genes like *CYP2C19*, there are hundreds of possible diplotypes, yet a relatively small number of possible metabolizer phenotypes listed previously. CPIC's standardized phenotype terms are included in standardized vocabularies such as Logical Observation Identifiers Names and Codes (LOINC) and can facilitate the use of clinical decision support tools, which can be triggered based on specific pharmacogenetic high-risk phenotypes. The CPIC *CYP2D6* Genotype to Phenotype Standardization Project worked to harmonize the assignment of *CYP2D6* metabolizer phenotypes from specific diplotypes in PGx test reports and clinical guidelines from CPIC and DPWG (Caudle et al., 2019).

As stated above, for some genes, CPIC uses an "activity value" (AV) to describe a graded activity for alleles. When the activity values of the maternal and paternal alleles are summed, the resulting "activity score" (AS) is used to map to guideline recommendations.

The diplotype–phenotype table also contains gene-based example consultation text to be provided with PGx test results, EHR priority result notation, and gene implementation workflows for EHR priority results.

## Allele frequency table

CPIC allele frequency tables summarize population-based allele frequencies reported in the literature. Studies are included in this table if the ethnicity of the population was clearly indicated, the genotype methodology is indicated, and the study represented an original publication. The ethnicities and locations reported in the articles are mapped into seven geographically defined groups (American, Central/South Asian, East Asian, European, Near Eastern, Oceanian, and Sub-Saharan African) and two admixed groups (African American/Afro-Caribbean and Latino), using the biogeographical grouping system developed by PharmGKB (Huddart, Fohner, et al., 2019a).

The allele frequency table contains multiple tabs summarizing allele frequencies by biogeographical group, diplotype frequencies by biogeographical group, phenotype frequency, and references; the latter describes allele frequencies for each publication included based on the criteria listed above.

There are several limitations regarding the accuracy of allele frequencies. Frequencies are based on published data which are limited or unavailable for some populations and many alleles and most studies test for a limited number of allelic variants that may lead to an underestimation of certain alleles. For example, c.430C > T (p.R144C) is often defaulted to a *CYP2C9*2* assignment, although this SNV is also present on *CYP2C9*35* and *CYP2C9*61*. Likewise, if no SNVs are found, *CYP2C9*1* is assigned, which inflates the frequency of this allele. All calculations based on allele frequencies from the primary literature are estimates at best and should be used with caution. Absence of a reported allele frequency does not necessarily indicate absence of an allele in that population and does not rule out that someone in that population might have the allele. It is important to keep in mind that the allele frequencies for each biogeographical group are averages of aggregated allele frequencies from multiple publications, each reporting on smaller, more specific study populations. The allele frequencies within each biogeographical group can range widely depending on the specific study population. Allele frequency tables are reviewed and updated as necessary each time guidelines are created or updated for the gene are published.

## Pre- and post-test alerts

CPIC provides files that contain drug- and gene-triggered workflows with example CDS text for pre- and post-test alerts.

## PharmCAT

The Pharmacogenomic Clinical Annotation Tool (PharmCAT; https://pharmcat.org) is a freely and publicly available software tool that extracts guideline variants from a genetic dataset represented as a variant call format (VCF) file, infers haplotypes and diplotypes, predicts drug phenotypes, and generates a report with CPIC genotype-based prescribing recommendations which can be used to inform treatment decisions (Klein & Ritchie, 2018; Sangkuhl et al., 2020). One of the challenges in implementing pharmacogenomics is the representation of the information in clinical dosing guidelines, including star-allele haplotypes, and extraction of these variants and haplotypes from genetic datasets. While PharmVar, PharmGKB, and CPIC provide a plethora of information about pharmacogene haplotypes, diplotypes, phenotypes, and ultimately drug response in patients, none of them provide tools to completely connect that information together enabling a user to convert genotypes into potential prescribing action. The motivation for the creation of PharmCAT was to connect whole genome or exome sequencing or genotypes to information from guideline recommendations in an automated way. The tool is accessible on GitHub under a Creative Commons license and an overview of the workflow is presented in Fig. 2.11.

PharmCAT takes a user provided normalized VCF file as input in conjunction with information packaged with the tool, which includes allele definition files and mapping information from haplotype to recommendations from CPIC and PharmGKB. PharmCAT has specific requirements for the VCF files: data should be provided using the GRCh38 reference sequence and all variant positions in the definition files should be reported in the file, including reference alleles. PharmCAT assumes missing positions are unknown and does not considered them for the genotype calls.

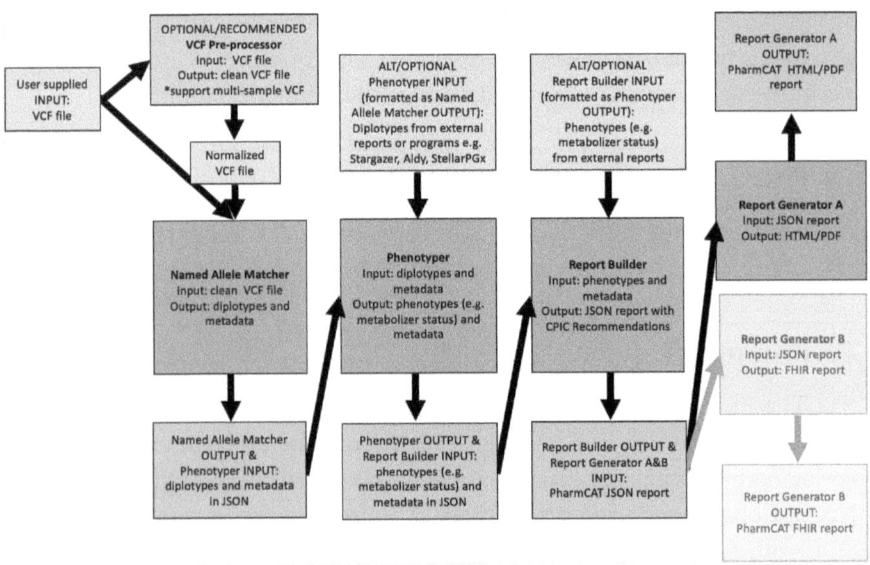


**Figure 2.11** The PharmCAT workflow. The PharmCAT workflow is split into modules, each is responsible for a specific task: diplotype assignment, phenotype prediction, recommendation mapping, and report generation. Report Generator B (faded boxes), which generates a report for electronic health record integration, is under development. The optional pre-processor helps users prepare VCF files for PharmCAT use.

Documentation is provided to guide users on how to process and normalize files for PharmCAT, and an optional preprocessor is available to aid the process. The entered VCF file is then compared to the allele definitions to generate the gene haplotype calls and diplotype assignments.

PharmCAT can process both phased and unphased data. In the case of unphased data, which most genotyping data are, more potential matches are possible. Potential matches are scored (more information at (https://github.com/PharmGKB/PharmCAT/wiki/NamedAlleleMatcher-101; https://github.com/PharmGKB/PharmCAT/wiki/NamedAlleleMatcher-201) and the top match is chosen, however the PharmCAT report provides warnings such as "The assigned diplotype is *1/*3A; however, *3B/*3C cannot be ruled out with unphased data." Following diplotype assignment, PharmCAT uses CPIC mappings to predict drug response phenotype which is used in turn to provide CPIC prescribing recommendations from CPIC's published guidelines.

Users also have the option of providing PharmCAT with externally sourced gene diplotypes, drug response phenotypes, or both, and if provided, the tool will use that information for recommendations in the final report. For *CYP2D6*, diplotypes cannot be accurately assigned based on pattern matching due to possible structural variation or copy number which cannot be readily represented in VCF format. *HLA* gene variation can be so complex that genotyping is not sufficient for allele assignment. In these cases, there are either specialized genetic tests which can be performed, specialized software tools designed for test interpretation, or both. The diplotype calls from those processes can also be uploaded to PharmCAT for phenotype prediction and mapping to guideline recommendations in the final report.

The PharmCAT report is an HTML document consisting of four parts: 1) genotype summary table, 2) CPIC recommendations section by drug in alphabetical order, 3) gene information about the interrogated variants, and 4) disclaimers. The length of the report is required in order to be fully transparent about each step of the process taken to generate the prescribing recommendations and to include all stipulations and warnings. For users who want a quick overview of the results, the first and second sections will suffice. The final two sections of the report are for those users who want the specifics of which (if any) alleles are not possible to determine given the type and quality of the data in the VCF input file.

The summary table lists genotypes and diplotypes by gene, corresponding drugs with guidelines, and resulting allele functionalities and phenotypes using CPIC standardized terms. The summary table also indicates if variant information is missing from the VCF (which could potentially result in changes to the predicted phenotype and/or CPIC recommendation). The second part of the report provides the diplotype-based CPIC recommendations by drug guideline with links to the CPIC website and guideline citations. The third section of the report summarizes the variants by gene found in the VCF input file. For each variant, the chromosomal position, rsID, and nucleotide calls are given, along with a list of alleles with definitions that include that variant. This section further highlights missing variant information and clearly indicates which alleles are affected by it. The fourth section lists limitations of the data and of the PharmCAT tool.

The HTML report is human readable, exportable as a PDF file and suitable for patients, clinicians, and scientists, but is not readily parsed or machine-readable. For this reason, PharmCAT is under development to generate results in Fast Healthcare Interoperability Resources (FHIR) format for easier integration into an EHR system. In addition, JavaScript Object Notation (JSON) output is available at various points of the PharmCAT workflow. This output is machine-readable and provides users with the flexibility to use portions of PharmCAT, such as the predicted diplotypes and phenotypes by gene, without parsing through the entire report.

## Conclusion

Multiple resources exist for cataloging, curating, synthesizing, and disseminating pharmacogenomic information. In addition to serving as public sources of information, these resources provide other pharmacogenomic databases, associations, and organizations with the critical information necessary to synthesize knowledge from primary and secondary sources of scientific evidence. Moreover, newly developed tools can utilize these resources to begin to remove some of the onus of synthesizing pharmacogenomic evidence from clinicians, patients, and researchers. Though each resource is valuable in itself, multiple resources can be a source of confusion for the pharmacogenomics community who are not always sure where to find each type of information. Additionally, though resource sharing and cross-communication is common for the projects discussed here, much pharmacogenomic primary data remain in silos. Ultimately, a single entry point for the pharmacogenomics community to find both knowledge and data would benefit both research and implementation efforts. Taking this idea even further, pharmacogenomic information eventually needs to be integrated with disease-centered genomics to realize its full potential as a critical component of personalized medicine. Clinical decision support systems that support both pharmacogenomics and disease genomics will enable patients to benefit from all areas of genomic medicine.

## References

Appell, M. L., Berg, J., Duley, J., Evans, W. E., Kennedy, M. A., Lennard, L., Marinaki, T., McLeod, H. L., Relling, M. V., Schaeffeler, E., Schwab, M., Weinshilboum, R., Yeoh, A. E. J., McDonagh, E. M., Hebert, J. M., Klein, T. E., & Coulthard, S. A. (2013). Nomenclature for alleles of the thiopurine methyltransferase gene. *Pharmacogenetics and Genomics, 23*(4), 242–248. https://doi.org/10.1097/FPC.0b013e32835f1cc0

Ashley, E. A., Butte, A. J., Wheeler, M. T., Chen, R., Klein, T. E., Dewey, F. E., Dudley, J. T., Ormond, K. E., Pavlovic, A., Morgan, A. A., Pushkarev, D., Neff, N. F., Hudgins, L., Gong, L., Hodges, L. M., Berlin, D. S., Thorn, C. F., Sangkuhl, K., Hebert, J. M., … Altman, R. B. (2010). Clinical assessment incorporating a personal genome. *The Lancet, 375*(9725), 1525–1535. https://doi.org/10.1016/S0140-6736(10)60452-7

Caudle, Kelly E., Dunnenberger, H. M., Freimuth, R. R., Peterson, J. F., Burlison, J. D., Whirl-Carrillo, M., Scott, S. A., Rehm, H. L., Williams, M. S., Klein, T. E., Relling, M. V., & Hoffman, J. M. (2017). Standardizing terms for clinical pharmacogenetic test results: Consensus terms from the clinical pharmacogenetics implementation consortium (CPIC). *Genetics in Medicine*, 215–223. https://doi.org/10.1038/gim.2016.87

Caudle, K. E., Klein, T. E., Hoffman, J. M., Müller, D. J., Whirl-Carrillo, M., Gong, L., McDonagh, E. M., Sangkuhl, K., Thorn, C. F., Schwab, M., Agúndez, J. A. G., Freimuth, R. R., Huser, V., Lee, M. T. M., Iwuchukwu, O. F., Crews, K. R., Scott, S. A., Wadelius, M., Swen, J. J., … Johnson, S. G. (2014). Incorporation of pharmacogenomics into routine clinical practice: The clinical pharmacogenetics implementation consortium (CPIC) guideline development process. *Current Drug Metabolism, 15*(2), 209–217. https://doi.org/10.2174/1389200215666140130124910

Caudle, K., Sangkuhl, K., Whirl-Carrillo, Swen, Haidar, C., Klein, T., Gammal, R., Relling, M., Scott, S., Hertz, D., Guchelaar, H., & Gaedigk, A. (2019). Standardizing CYP2D6 genotype to phenotype translation: Consensus recommendations from the clinical pharmacogenetics implementation consortium and Dutch pharmacogenetics working group. *Clinical and Translational Science, 13*(1). https://doi.org/10.1111/cts.12692. Epub

Chen, R., Mias, G. I., Li-Pook-Than, J., Jiang, L., Lam, H. Y. K., Chen, R., Miriami, E., Karczewski, K. J., Hariharan, M., Dewey, F. E., Cheng, Y., Clark, M. J., Im, H., Habegger, L., Balasubramanian, S., O'Huallachain, M., Dudley, J. T., Hillenmeyer, S., Haraksingh, R., … Snyder, M. (2012). Personal omics profiling reveals dynamic molecular and medical phenotypes. *Cell, 148*(6), 1293–1307. https://doi.org/10.1016/j.cell.2012.02.009

*Clinical pharmacogenomics: Premarket evaluation in early-phase clinical studies and recommendations for labeling.*(2013). https://www.fda.gov/regulatory-information/search-fda-guidance-documents/clinical-pharmacogenomics-premarket-evaluation-early-phase-clinical-studies-and-recommendations.

Daly, A. K., Brockmöller, J., Broly, F., Eichelbaum, M., Evans, W. E., Gonzalez, F. J., Huang, J. D., Idle, J. R., Ingelman-Sundberg, M., Ishizaki, T., Jacqz-Aigrain, E., Meyer, U. A., Nebert, D. W., Steen, V. M., Wolf, C. R., & Zanger, U. M. (1996). Nomenclature for human CYP2D6 alleles. *Pharmacogenetics, 6*(3), 193–200. https://doi.org/10.1097/00008571-199606000-00001

Demir, E., Cary, M. P., Paley, S., Fukuda, K., Lemer, C., Vastrik, I., Wu, G., D'Eustachio, P., Schaefer, C., Luciano, J., Schacherer, F., Martinez-Flores, I., Hu, Z., Jimenez-Jacinto, V., Joshi-Tope, G., Kandasamy, K., Lopez-Fuentes, A. C., Mi, H., Pichler, E., … Bader, G. D. (2010). The BioPAX community standard for pathway data sharing. *Nature Biotechnology, 28*(9), 935–942. https://doi.org/10.1038/nbt.1666

Dewey, F. E., Chen, R., Cordero, S. P., Ormond, K. E., Caleshu, C., Karczewski, K. J., Whirl-Carrillo, M., Wheeler, M. T., Dudley, J. T., Byrnes, J. K., Cornejo, O. E., Knowles, J. W., Woon, M., Sangkuhl, K., Gong, L., Thorn, C. F., Hebert, J. M., Capriotti, E., David, S. P., … Ashley, E. A. (2011). Phased whole-genome genetic risk in a family quartet using a major allele reference sequence. *PLoS Genetics,* 7(9). https://doi.org/10.1371/journal.pgen.1002280

Dewey, F. E., Grove, M. E., Pan, C., Goldstein, B. A., Bernstein, J. A., Chaib, H., Merker, J. D., Goldfeder, R. L., Enns, G. M., David, S. P., Pakdaman, N., Ormond, K. E., Caleshu, C., Kingham, K., Klein, T. E., Whirl-Carrillo, M., Sakamoto, K., Wheeler, M. T., Butte, A. J., … Quertermous, T. (2014). Clinical interpretation and implications of whole-genome sequencing. *JAMA - Journal of the American Medical Association, 311*(10), 1035–1044. https://doi.org/10.1001/jama.2014.1717

Gaedigk, A., Ingelman-Sundberg, M., Miller, N. A., Leeder, J. S., Whirl-Carrillo, M., & Klein, T. E. (2018). The pharmacogene variation (PharmVar) consortium: Incorporation of the human cytochrome P450 (CYP) allele nomenclature database. *Clinical Pharmacology and Therapeutics, 103*(3), 399–401. https://doi.org/10.1002/cpt.910

Gaedigk, A., Sangkuhl, K., Whirl-Carrillo, M., Twist, G. P., Klein, T. E., & Miller, N. A. (2019). The evolution of PharmVar. *Clinical Pharmacology and Therapeutics, 105*(1), 29–32. https://doi.org/10.1002/cpt.1275

Gaedigk, A., Simon, S., Pearce, R., Bradford, L., Kennedy, M., & Leeder, J. (2007). The CYP2D6 activity score: Translating genotype information into a qualitative measure of phenotype. *Clinical Pharmacology and Therapeutics, 83*(2), 234–242. https://doi.org/10.1038/sj.clpt.6100406

Huddart, R., Fohner, A. E., Whirl-Carrillo, M., Wojcik, G. L., Gignoux, C. R., Popejoy, A. B., Bustamante, C. D., Altman, R. B., & Klein, T. E. (2019). Standardized biogeographic grouping system for annotating populations in pharmacogenetic research. *Clinical Pharmacology and Therapeutics, 105*(5), 1256–1262. https://doi.org/10.1002/cpt.1322

Huddart, R., Sangkuhl, K., Whirl-Carrillo, M., & Klein, T. E. (2019). Are randomized controlled trials necessary to establish the value of implementing pharmacogenomics in the clinic? *Clinical Pharmacology and Therapeutics, 106*(2), 284–286. https://doi.org/10.1002/cpt.1420

Ingelman-Sundberg, M., Daly, A. K., Oscarson, M., & Nebert, D. W. (2000). Human cytochrome P450 (CYP) genes: Recommendations for the nomenclature of alleles. *Pharmacogenetics, 10*(1), 91–93. https://doi.org/10.1097/00008571-200002000-00012

Jeiziner, C., Suter, K., Wernli, U., Barbarino, J. M., Gong, L., Whirl-Carrillo, M., Klein, T. E., Szucs, T. D., Hersberger, K. E., & Meyer zu Schwabedissen, H. E. (2021). Pharmacogenetic information in Swiss drug labels – a systematic analysis. *The Pharmacogenomics Journal, 21*(4), 423–434. https://doi.org/10.1038/s41397-020-00195-4

Johnson, J. A., Burkley, B. M., Langaee, T. Y., Clare-Salzler, M. J., Klein, T. E., & Altman, R. B. (2012). Implementing personalized medicine: Development of a cost-effective customized pharmacogenetics genotyping array. *Clinical Pharmacology and Therapeutics, 92*(4), 437–439. https://doi.org/10.1038/clpt.2012.125

Klein, T. E., & Ritchie, M. D. (2018). PharmCAT: A pharmacogenomics clinical annotation tool. *Clinical Pharmacology and Therapeutics, 104*(1), 19–22. https://doi.org/10.1002/cpt.928

Lever, J., Barbarino, J. M., Gong, L., Huddart, R., Sangkuhl, K., Whaley, R., Whirl-Carrillo, M., Woon, M., Klein, T. E., & Altman, R. B. (2020). PGxMine: Text mining for curation of pharmgKB. In *Pacific symposium on biocomputing* (Vol. 25, pp. 611−622). World Scientific Publishing Co. Pte Ltd. Issue 2020 http://psb.stanford.edu/.

Oscarson, M., & Ingelman-Sundberg, M. (2002). CYPalleles: A web page for nomenclature of human cytochrome P450 alleles. *Drug Metabolism and Pharmacokinetics, 17*(6), 491−495. https://doi.org/10.2133/dmpk.17.491

Relling, M. V., & Klein, T. E. (2011). CPIC: Clinical pharmacogenetics implementation consortium of the pharmacogenomics research network. *Clinical Pharmacology and Therapeutics, 89*(3), 464−467. https://doi.org/10.1038/clpt.2010.279

Relling, Mary V., Klein, T. E., Gammal, R. S., Whirl-Carrillo, M., Hoffman, J. M., & Caudle, K. E. (2019). The clinical pharmacogenetics implementation consortium: 10 Years later. *Clinical Pharmacology and Therapeutics, 107*(1), 171−175. https://doi.org/10.1002/cpt.1651

Sangkuhl, K., Whirl-Carrillo, M., Whaley, R. M., Woon, M., Lavertu, A., Altman, R. B., Carter, L., Verma, A., Ritchie, M. D., & Klein, T. E. (2020). Pharmacogenomics clinical annotation tool (PharmCAT). *Clinical Pharmacology and Therapeutics, 107*(1), 203−210. https://doi.org/10.1002/cpt.1568

Shows, T. B., McAlpine, P. J., Boucheix, C., Collins, F. S., Conneally, P. M., Frézal, J., Gershowitz, H., Goodfellow, P. N., & Hall, J. G. (1987). Guidelines for human gene nomenclature. An international system for human gene nomenclature (ISGN, 1987). *Cytogenetics and Cell Genetics, 46*(1−4), 11−28.

Swen, J., Nijenhuis, Boer, Zee, a L., Maitland-van der, Mulder, H., Rongen, G., Schaik, Schalekamp, T., Touw, Weide, J., van der, Wilffert, B., Deneer, V., & Guchelaar, H. J. (2011). Pharmacogenetics: From bench to byte–an update of guidelines. *Clinical Pharmacology and Therapeutics, 89*(5). https://doi.org/10.1038/clpt.2011.34. Epub

Swen, J. J., Wilting, I., Goede, A. D., Grandia, L., Mulder, H., Touw, D. J., Boer, A. D., Conemans, J. M. H., Egberts, T. C. G., Klungel, O. H., Koopmans, R., Weide, J. V. D., Wilffert, B., Guchelaar, H. J., & Deneer, V. H. M. (2008). Pharmacogenetics: From bench to byte. *Clinical Pharmacology and Therapeutics, 83*(5), 781−787. https://doi.org/10.1038/sj.clpt.6100507

Twist, G. P., Gaedigk, A., Miller, N. A., Farrow, E. G., Willig, L. K., Dinwiddie, D. L., Petrikin, J. E., Soden, S. E., Herd, S., Gibson, M., Cakici, J. A., Riffel, A. K., Leeder, J. S., Dinakarpandian, D., & Kingsmore, S. F. (2017). Erratum: Constellation: A tool for rapid, automated phenotype assignment of a highly polymorphic pharmacogene, CYP2D6, from whole-genome sequences. *Npj Genomic Medicine, 2*, 16039. https://doi.org/10.1038/npjgenmed.2016.39

Vatsis, K. P., Webe, W. W., Bell, D. A., Dupret, J. M., Evans, D. A. P., Grant, D. M., Hein, D. W., Lin, H. J., Meyer, U. A., Relling, M. V., Sim, E., Suzuki, T., & Yamazoe, Y. (1995). Nomenclature for N-acetyltransferases. *Pharmacogenetics, 5*(1), 1−17. https://doi.org/10.1097/00008571-199502000-00001

Whirl-Carrillo, M., Huddart, R., Gong, L., Sangkuhl, K., Thorn, C. F., Whaley, R., & Klein, T. E. (2021). An evidence-based framework for evaluating pharmacogenomics knowledge for personalized medicine. *Clinical Pharmacology and Therapeutics, 110*(3), 563−572. https://doi.org/10.1002/cpt.2350

Whirl-Carrillo, M., McDonagh, E. M., Hebert, J. M., Gong, L., Sangkuhl, K., Thorn, C. F., Altman, R. B., & Klein, T. E. (2012). Pharmacogenomics knowledge for personalized medicine. *Clinical Pharmacology and Therapeutics, 92*(4), 414−417. https://doi.org/10.1038/clpt.2012.96

CHAPTER THREE

# Laboratory considerations for pharmacogenomic testing

**Debbie M. Figueroa[1], Ender Karaca[1], Blake Atwood[1,2], Michele Erickson-Johnson[1,2]**
[1]Sanford Health Imagenetics, Sanford Medical Genetics Laboratory, Sioux Falls, SD, United States
[2]University of South Dakota Sanford School of Medicine, Sioux Falls, SD, United States

## Contents



## Introduction

Clinical laboratory considerations are essential to successfully developing and implementing CDS for pharmacogenetics and precision medicine. This chapter presents components needed to establish a clinical molecular diagnostics laboratory specializing in pharmacogenetics. Elements integral to the setup and certification of a clinical molecular lab such as regulatory bodies, resources, personnel, and assay validation are discussed. Approaches used to correlate and phase genotypes to designate star allele and metabolizer status calls and considerations when determining what variants to test are then presented. An overview of specimen tracking, type, and

*Clinical Decision Support for Pharmacogenomic Precision Medicine*
ISBN: 978-0-12-824453-1
https://doi.org/10.1016/B978-0-12-824453-1.00006-3

quality assessment is provided. Testing platforms used for genotyping and copy number assessment are compared, including their strengths and weaknesses. Overall, this chapter aims to provide a comprehensive overview of the regulatory and testing components that merit careful consideration when establishing a pharmacogenomics clinical molecular diagnostics laboratory. Key abbreviations used in this chapter are summarized in Box 3.1.

**Box 3.1 Key abbreviations and definitions**

AOH—Absence of heterozygosity
CAP—College of American Pathologists
CDC—Centers for Disease Control and Prevention
CDS—Clinical decision support
CLIA—Clinical Laboratory Improvement Amendment of 1988
CLSI—Clinical Laboratory Standards Institute
CMS—Centers of Medicare & Medicaid
CPIC—Clinical Pharmacogenetics Implementation Consortium
DPWG—Dutch Pharmacogenetics Working Group
CN—Copy Number
CNV—Copy Number Variation
EMR—Electronic medical record system
FDA—Food and Drug Administration
HGVS—Human Genome Variation Society
InDel—Insertion or Deletion (small sequence length changes in a target region)
IT—Information Technology
LDT—Laboratory developed test
LIMS—Laboratory information management system
LRS—Long-read sequencing
MLPA—Multiplex Ligation-dependent Probe Amplification
NGS—Next-Generation Sequencing
qPCR—Quantitative Polymerase Chain Reaction
PCR—Polymerase Chain Reaction
PGx—Pharmacogenetics
PPM—Provider Performed Microscopy Procedures
PRS—Polygenic Risk Score
SBS—Sequencing By Synthesis
SMRT—Single molecule real time sequencing technology
SNP—Single Nucleotide Polymorphism
SRS—Short-read sequencing
TAT—Turnaround time
TJC—The Joint Commission
TGS—Third-Generation Sequencing, long-read sequencing
WES—Whole exome sequencing
WGS—Whole genome sequencing

## Clinical laboratory regulation

Any laboratory performing pharmacogenomic testing should meet all requirements set forth by federal, state, and local governments. The Centers for Medicare and Medicaid Services (CMS) regulates all laboratory testing, regardless of complexity, through the Clinical Laboratory Improvement Amendments (CLIA). CLIA sets forth regulations that include minimum standards that apply to all laboratories within the United States (US) "… that test human specimens for health assessment or to diagnose, prevent, or treat disease" ((CLIA), 2021). CLIA categorizes laboratories into waived, moderately complex and highly complex laboratories, the latter of which corresponds to pharmacogenomic testing. Each of these categories has slightly different regulatory requirements based on the complexity of the testing performed. Any lab performing nonwaived testing is subject to inspection and must meet, at minimum, the CLIA requirements ((CLIA), 2021).

Prior to testing, the appropriate CLIA certificate must be obtained (Table 3.1). Pharmacogenomic testing requires, at minimum, a Certificate of Compliance. Although the Certificate of Accreditation is voluntary, it communicates to providers, patients, and others in the industry that the laboratory is dedicated to providing the highest quality laboratory testing and holds the laboratory accountable to a published set of standards and best practices.

The laboratory is also subject to accreditation requirements. There are several accrediting bodies recognized by CMS (Table 3.2). The College of American Pathologists (CAP) and The Joint Commission (TJC) support clinical pathology and anatomic pathology type labs, such as those that might provide pharmacogenomic or genetic testing.

In addition to CLIA and Accreditation standards, laboratories need to consider state-specific requirements for in-state and out-of-state samples, if applicable. In addition, while accreditation programs provide checklists that are specific to the type of laboratory testing being performed, new laboratories may not know where to start. There are many resources available that provide a quality framework and provide a level of awareness in managing, monitoring, and maintaining quality lab practice, such as those provided by the Clinical Laboratory Standards Institute (CLSI; https//clsi.org). CLSI is a global community of laboratory experts that provides a variety of training programs, education resources, and documents that aid laboratories in meeting and maintaining quality management systems and best practices.

**Table 3.1** Clinical laboratory improvements amendment certificate categories.

| Certificate | Description |
|---|---|
| Certificate of Waiver | Certificate of Waiver only allows laboratories to perform waived, and unmodified FDA approved tests. These laboratories are not usually subjected to routine inspections. |
| Certificate of Provider Performed Microscopy Procedures (PPM) | Certificate of PPM is appropriate for laboratories that perform a limited test menu during a patient's visit with a healthcare provider, examples include wet mounts, pinworm examinations, and potassium hydroxide (KOH) preparations. These procedures require a licensed physician, midlevel practitioner/ advanced practice provider, or dentist. |
| Certificate of Registration | The Certificate of Registration allows moderate or high complexity labs to operate until a CLIA survey or accreditation inspection is complete. |
| Certificate of Compliance<br>Certificate of Accreditation | The Certificates of Compliance or of Accreditation replace the Certificate of Registration when a lab satisfies the minimum requirements of a CLIA survey or after a successful accreditation inspection, respectively. |

This table describes five different CLIA certificate categories into which laboratories fall.

**Table 3.2** List of accrediting bodies for laboratories.

| List of accrediting bodies for clinical laboratories |
|---|
| The College of American Pathologists (CAP) |
| The Joint Commission (TJC) |
| Commission on Office Laboratory Accreditation (COLA) |
| American Association for Laboratory Accreditation |
| American Association of Blood Banks (AABB) |
| Accreditation Association for Hospitals and Health Systems/Healthcare facilities Accreditation Program (AAHHS/HFAP)[a] |
| American Society for Histocompatibility and Immunogenetics (ASHI) |

[a]Formerly known as the American Osteopathic Association/Healthcare Facilities Accreditation Program (AOA/HFAP).

## Laboratory personnel considerations

CLIA regulations require that high-complexity laboratories have qualified individuals performing testing. At a minimum, such laboratories must have a qualified director, clinical consultant, technical supervisor/section director, general supervisor, and testing personnel/laboratory technicians. The laboratory director can function in more than one role and can delegate some duties to others that meet the qualifications as specified by CLIA. Other positions that could be important for a clinical molecular laboratory include a research and development (R&D) team to validate new or expanded testing methods; a bioinformatics team if haplotyping will be called in-house or if Next-Generation Sequencing (NGS) data will be used for genotyping; or an information technology (IT) support team to incorporate and maintain analysis software, reporting tools, and ensure patient health information is maintained and secure. Finally, many laboratories also staff genetic counselors to assist with clinical case review and ensure test appropriateness, draft results reports, return results to clients, and respond to clinical questions from the patient or provider. Not all of these positions are required, and in some cases an experienced laboratory director may be able to fill most of these roles.

Laboratories may also rely on external vendors for many aspects of pharmacogenomic testing and reporting, which can potentially cost more or less than in-house services. While less effort is required of the institution if all roles are not filled internally, it often means the institution is fully dependent on the vendor. This can be problematic if a vendor can no longer stay in business or makes unwanted changes, which could seriously disrupt the pharmacogenomics program of the institution.

## Assay validation considerations

The laboratory director must also review and approve that clinical testing has been properly validated or verified prior to offering a clinical test. CLIA, CAP, and others define the minimum validation requirements for modified FDA-Cleared/Approved tests and LDTs and additional guidance is available from CLSI (Quality Management for Molecular Genetic Testing; Approved Guideline. CLSI Document MM20-A (ISBN 1-56238-859-2), n.d.). Minimum requirements for validation studies must define specimen types (blood, saliva, buccal swabs, etc.), analytical accuracy, analytical precision/reproducibility, reportable range, analytical sensitivity (lower detection limit), analytical specificity, and reference intervals (Table 3.3). Clinical performance characteristics describe the diagnostic sensitivity and specificity, positive and negative predictive values or likelihood ratios, and clinical utility. Clinical utility of an assay/test is commonly referred to as results that are used in the clinical decision-making process. Data from the literature can be used to support clinical performance characteristics. However, labs should

**Table 3.3** Key validation requirements defined.
*Important validation requirements for clinical laboratories.*

| Term | Definition | Quantitative assays | Qualitative assays |
|---|---|---|---|
| Analytical Accuracy | Closeness of the agreement between the result of a measurement and a true value. Compare results to a definitive or reference method using well-characterized reference material. For genetic tests with limited number of genotypes the lab should verify the ability of the assay to detect genotypes. Typical acceptance criteria is 100%. | "Closeness to true." A minimum of 20* samples should be used distributed across the Analytical Measurement Range (AMR). | Correlation to a comparative test to establish "true." A minimum of 20* samples should be used including positive, negative, and low-positive with concentrations approaching the limit of detection, when applicable. |
| Analytical Precision/ Reproducibility | Shows that the test returns the same result regardless to minor variation in testing; i.e., 3 × 3 × 3 matrix, three different technologists, three samples in triplicate (intraassay, i.e., within-run and interassay, i.e., between-run). | Calculate a mean and coefficient of variation (CV). | Shows results are concordant. |
| Reportable Range | Full range of reported values. For sequencing assays the location and which parts of the gene to be sequenced should be noted. The validation should also describe expected results (normal, heterozygous, or mutant). | Analytical Measurement Range (AMR)—calibration standards. | Homozygous wild type, heterozygous, or homozygous variant. |

| | | | |
|---|---|---|---|
| Analytical Sensitivity | Lower limit of detection. Ability of the assay to detect a minor allele or a variant in the background of other pathogens, rare variants, chimerism, mosaicism, or tumor-normal mixtures. | – | Sequencing assays should establish the minimum amount of DNA that will be acceptable. Some assay types may require validation of minimum amounts of detectable mutant DNA in a background of normal. |
| Analytical Specificity | Ability of a test to correctly identify or quantity in the presence of interfering or cross-reactive substances. | – | Validation plans should include documentation that interference by primers by SNPs or mutations is not expected. Generally several different databases should be searched for known SNPs and mutations. |
| Reference Intervals | The range of test values expected for a designated population of persons [i.e., 95% of persons that are presumed to be healthy (or normal)]. | – | Reference interval may include all genotypes or if the value depends on a clinical situation then a plan for interpretation of patient results must be defined (MOL.31255). For example: A normal result for this test is "no mutation detected." The reference range will be defined using GenBank reference sequence NC_0000000. |

rely on clinical validation when possible. Other performance characteristics such as specimen and reagent stability, carryover, and cross-contamination must also be considered.

## Genetic terminology and nomenclature

In order for geneticists, pharmacists, and other healthcare providers to be able to communicate laboratory results for pharmacogenomic tests, it is helpful to understand the nomenclature for genetic variants. Genetic variants are identified and named following the Human Genome Variation Society (HGVS) guidelines (http://varnomen.hgvs.org/). The HGVS focuses on individual variants compared to star allele nomenclature, which describes alleles. HGVS nomenclature uses genomic location in reference to either the chromosome or the gene to indicate the exact variant named. For instance, the variant that constitutes the *4 allele in *CYP2D6* is located on chromosome 22. The genomic nomenclature for this variant is written as GRCh38 (NC_000022.11):g.42128945C > T where GRCh38 refers to the genome build, NC_000022.11 refers to the reference file used and indicates it is on chromosome 22, g.42128945 indicates the location along chromosome 22 where the variant is located and C > T indicates that the cytosine at this position has been changed to a thymine. This same variant can also be identified based on its location in the coding portion of the *CYP2D6* gene, NM_000106.6:c.506-1G > A. The NM_000106.6 refers to the reference file used for *CYP2D6* and the c.506−1 is the location of the variant relative to the coding portions of the gene. This variant has a −1 which indicates that this variant is in the intron one base before the nearest exon where nucleotide 506 is located (Fig. 3.1). In this case, G > A indicates that a guanine has been replaced with an adenine. This is different from the genomic location which had this same variant as a C > T. This is because the *CYP2D6* gene is located on the antisense strand of the chromosome. Genomic positions always reference the sense strand, whereas genes can be found on either the sense or antisense strand of the chromosome. For more information about genetic nomenclature, please see HGVS (HYPERLINK "http://varnomen.hgvs.org/" \o "http://varnomen.hgvs.org/"http://varnomen.hgvs.org/).

Star alleles can be comprised of a single genetic variant or multiple genetic variants observed together on the same chromosome in a given gene. For instance CYP2D6*2 is a combination of GRCh38 (NC_000022.11): g.42127941G> A and g.42126611C > G (or NM_000106.6:c.886C > T

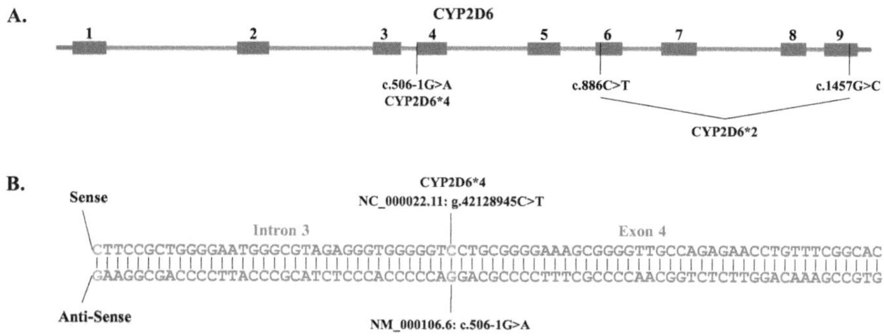


**Figure 3.1** CYP2D6 gene layout. *CYP2D6* is located on chromosome 22 and is composed of nine exons and eight introns. The *4 allele contains a single variant found in intron 3 (A). The *2 allele is comprised of two variants (c.886C > T and c.1457G > C) found in exons 6 and 9. The *4 allele location is the last base in intron 3 (orange sequence) at the border of exon 4 (green sequence) (B). If referring to the genomic coordinates the *4 defining variant is a C > T since it is located on the upper or "sense" strand. If referring to the coding DNA sequence, the variant is G > A since *CYP2D6* transcribed or "coded" by the bottom or "antisense" strand.

and c.1457G > C using the coding sequence) where both variants must be present on the same chromosome to call the *2 allele (Fig. 3.1). Thus, to call a phenotype all genotype variants must be taken into account.

A challenge that all labs face is accurately calling the phenotype based on the genotypes present. Most phenotype calls are straightforward when a single heterozygous or homozygous variant is present in the gene of interest. Issues arise when there are two or more heterozygous variants or copy number variants (CNVs) in the gene (Nofziger & Paulmichl, 2018). The challenge is determining if the variants are located on the same or opposite chromosomes, also known as the phase of the variants. If the phasing is unclear, then laboratories need to make assumptions based on what is known about the gene, also known as inference. Workgroups have created recommendations to standardize this process for heterozygous variants that have been repeatedly observed on the same allele (i.e., "in phase") due to linkage disequilibrium (Kalman et al., 2016). As will be discussed, different tools may be used to assign the star alleles for a gene based on the variants detected with relative accuracy (Yang et al., 2017). These tools make imputations based on published literature, frequency in the general population, and expert workgroup recommendations. Importantly, genotypes can be correctly phased using these tools in the majority of cases, but there is a risk that the wrong phenotype could be called in this way.

## Determining genes and variants to test

Prior to implementing a pharmacogenomic test, the decision must be made as to which genes and variants are included. Most institutions rely on published guidelines from professional groups and consortiums such as the Clinical Pharmacogenetics Implementation Consortium (CPIC) and Dutch Pharmacogenetics Working Group (DPWG), and regulatory guidance from the FDA (Abdullah-Koolmees et al., 2021; Pratt et al., 2018, 2021). These groups utilize published evidence along with professional evaluation to provide guidance on which genes and variants to include on a pharmacogenomic panel, how to translate genotyping results, and which therapeutic recommendations to make for specific drugs. Other organizations rely on research, experience, and the knowledge base of their own experts. The most important factor is strong published scientific evidence demonstrating the genes and variants selected have a well-established effect on drug metabolism. Utilizing professional guidelines is strongly advised to maintain consistency in the field.

An example of a gene with a well-established interaction with a drug is *CYP2D6* and codeine. There are more than 300 published papers linking the function of the *CYP2D6* protein to the conversion of codeine to morphine. The evidence includes population studies (Neafsey et al., 2009), functional and enzyme kinetic studies (Hu et al., 2016), studies in multiple ethnic backgrounds (Brosen, 2015; Virbalas et al., 2019; Yue et al., 1995), differences between men and women (Lopes et al., 2020), in pregnant and breastfeeding women (Badaoui et al., 2021; Madadi et al., 2009), in children (Cartabuke et al., 2014; Ramsey et al., 2020; Williams et al., 2002), and even in animals (Knych et al., 2019). On the other hand, *NAT1* is an example of a gene that may not have enough information to date to make clinical decisions. There have been publications linking it to cotinine (Lemasters et al., 2015), thiopurine, and aminosalicylate (Stocco et al., 2015) metabolism, but there have not been sufficient follow-up studies to confirm these results nor to establish how it could be used clinically. Some publications evaluate the possible implications of different *NAT1* variants (Hein et al., 2018), but the breadth and depth of information is not sufficient for clinical use at this time.

Institutions should also be aware that not all pharmacogenomic variants are the same. For example, some are a single base change, such as an adenine replacing a cytosine at a particular site. This type of change can easily be assayed using a variety of methods. Other changes are more complex, such as nucleotide repeats and large deletions, duplications, or rearrangements of the DNA. These changes cannot always be easily detected and their detection often requires sophisticated molecular methods. For example, the promoter region of *UGT1A1* contains six TA repeats (TATATATATATA) in the normal state. An increase in the number of the TA repeats to seven (TATATATATATATA) leads to decreased expression of *UGT1A1* and thus reduced enzymatic activity (Marques & Ikediobi, 2010). It is more difficult to test for this type of change with a targeted genotyping assay, as such an alternative method, like sequencing, would be required.

When designing a pharmacogenomic testing strategy, there are two primary approaches. The first is to design a panel that includes only select variants with known effects on drug metabolism, in the pharmacogenes that will be reported. Depending on the testing method (i.e., genotyping vs. sequencing), such an approach can minimize the test cost since the panel will only include well-established variants with known gene–drug interactions. The drawback to this approach is that when a new guideline becomes available that recommends evaluating additional variants, further development and validation work are required to add new variants to the panel.

The second approach involves including as many genes and variants as possible in a large panel, regardless of whether they are supported by strong scientific evidence. This is typically better suited for NGS approaches as compared with a genotyping approach. This approach has advantages during the development stage since it requires a single large validation and may not require additional validations each time a new variant is added. When new guidelines become available, variants already included in the panel can be added to the reporting workflow after minor verifications are performed. The drawback to this approach is the extra cost of sequencing additional variants on each run that will not be reported initially. Thus, when selecting and building a panel, the laboratory needs to weigh the benefit of a more expensive and comprehensive panel that will potentially require fewer validation efforts and resources over time versus a less expensive genotyping panel, which includes current actionable variants, but will require more effort to include additional variants later on.

It is up to the discretion of any organization to determine the evidence threshold required for reporting each variant. Some organizations have opted to share variant information such as *APOE* on what they term an "educational" or "research" basis (i.e., not to guide clinical care) (Green et al., 2009). If organizations decide to report variants on an educational or research basis they must be aware that this information could still lead to clinical decisions.

Another consideration when designing a test is that some star alleles are based on haplotypes and not necessarily individual variants themselves. For

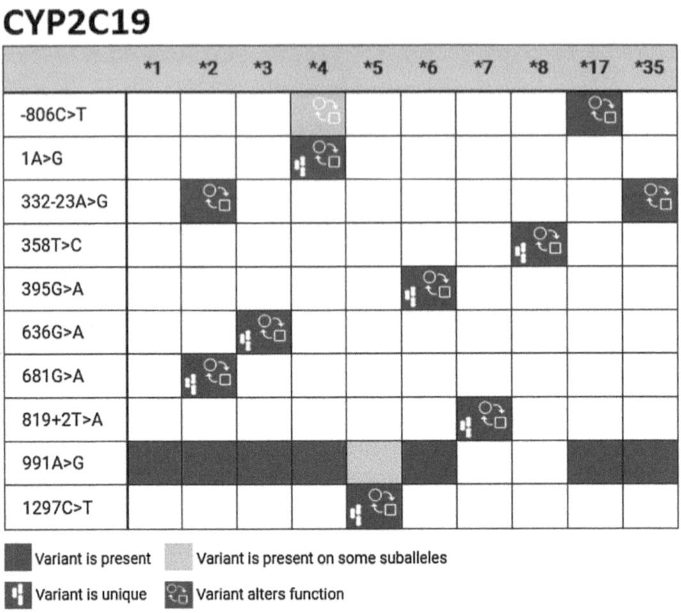


**Figure 3.2** Example star allele genotype comparison chart from PharmVar for CYP2C19. The c. nomenclature is based on the mRNA reference number NM_000769.1. *Used with permission from PharmVar (https://www.pharmvar.org/).*

instance, in *CYP2C19,* the c.−806C > T variant alters function of the protein and is the primary variant in the *17 allele but can also be present in the *4 allele (Fig. 3.2, see gray box under *4). The determining factor for this variant to be classified as *4 instead of *17 is whether the c.1A > G variant is also present. The presence of the c.1A > G variant with or without the c.−806C > T variant takes priority and will always be classified as *4 instead of *17 (Fig. 3.2). Other variants, like c.991A > G in *CYP2C19,* are present in several haplotypes but do not alter protein function. In some circumstances, these variants can be used to help infer the correct haplotype. Most institutes will not include this type of variant in the testing or reporting since it would increase the cost of testing and only add minimal benefit.

## Obtaining, tracking, and assessing DNA quality of samples

### Acceptable sample types

Whole blood and saliva are both commonly used samples for the assessment of constitutional variants in pharmacogenomics. Although not as common, skin biopsy or buccal swab may be alternative options, especially for patients that have undergone bone marrow transplant or have neoplasms that may result in low white blood cell count or high levels of somatic mosaicism (e.g., leukemic cell burden). Historically, extraction of high-quality DNA involved the use of buffered salt solutions, alcohol, and possibly phenol and chloroform. This process has become more standardized with the availability of kits that allow extraction by either manual or automated means. Several companies manufacture kits for isolation of DNA from whole blood, saliva, or other tissues.

## Sample tracking and management

According to the Centers for Disease Control, approximately 70% of medical decisions made in the United States depend on laboratory test results. Because of the importance of accurate testing and reporting, it is imperative that laboratories have a sample tracking system that provides effective and continuous management throughout all the sample processing steps. Clinical laboratories and biobanks use various samples, including peripheral blood, saliva, fresh tissue, and stabilized samples including formalin-fixed and paraffin-embedded tissue. Depending on the capacity, each laboratory processes dozens to hundreds of samples every day. A vast majority of these samples undergo several complex laboratory workflows that involve multiple manipulations by testing personnel, highlighting a potential risk for mislabeling and mishandling, which could result in severe sample result and reporting errors. Such systems ensure sample identity and integrity by supporting workflow and data tracking in a flexible architectural context. This is especially true in the case of pharmacogenomic tests. These may be ordered as standalone tests (e.g., single-gene test ordered in response to a specific medication) or as part of a larger screening panel in a healthy individual (e.g., a panel-based test to inform future drug therapy decisions. Results may need to be classified as clinical, educational, or research and samples may be stored beyond testing for research or other purposes. In addition, pharmacogenomic testing may be complex and involve multiple steps, such as with copy number analyses for *CYP2D6* gene.

Today, many clinical genetics laboratories achieve proper sample tracking and management using software-based *Laboratory Information Management Systems* (LIMS). Although commercial LIMS software is broadly available and employed by the majority of clinical laboratories, laboratories can create their own LIMS depending on their specific needs. Some electronic health record (EHR) companies also provide the LIMS component within the EHR software, which can streamline engagement and maintenance processes for IT and the laboratory.

Although assays and workflows across different laboratories are heterogenous, all LIMS have common core elements (Fig. 3.3) (Cucoranu, 2016). The first component is the "accessioning and processing" step, where the clinical sample is introduced to the LIMS. At this stage, samples are given an accession number and then registered in an electronic log sheet along with the sample type (e.g., blood, saliva, tissue) and amount, collection and receiving dates, and the patient's demographic data. Necessary clinical information and the indication for the ordered test are often recorded in this step. Many laboratories utilize a barcoding system at this step to facilitate accurate and efficient sample tracking throughout different assay workflows. Extracted DNA or RNA is then quantified, and results are recorded to ensure there is sufficient quantity for the planned assays. In the following step, samples are sorted and aliquoted according to each sample's assigned genetic test. If the assay is not run right away, samples should be stored at the appropriate temperature (e.g., in the refrigerator at 4°C or in the freezer at −20°C).

After the assays are run, data are exported to the LIMS for initial review. In this step, the integrity and quality of the data are evaluated by the testing personnel. Results that pass this initial check are then transferred for final review and reporting by the appropriately qualified laboratory director or delegates as appropriate. Depending on the functionality of the LIMS and the laboratory's desire, the LIMS can also incorporate a wide range of processes

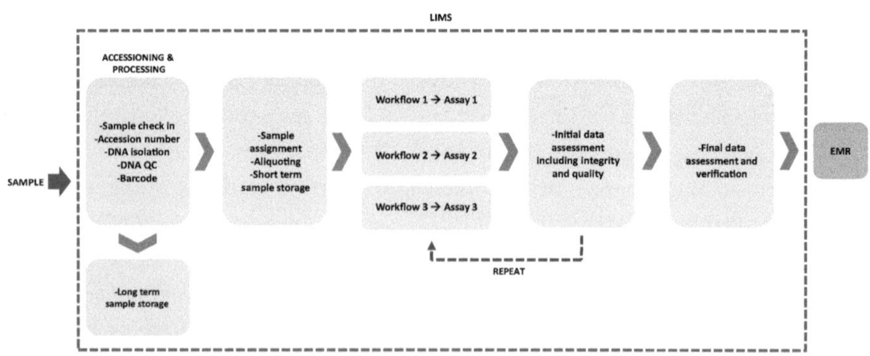


**Figure 3.3** Schematic diagram of a basic laboratory information management system (LIMS). Although assays and workflows across different laboratories are heterogenous, all LIMS have common core elements.

such as final verification, report creation, and data storage. Nevertheless, many laboratories employ separate reporting platforms, either as a component of the utilized EHR or a standalone tool.

The LIMS should be highly flexible to keep up with constant advances in technology that introduces new tools and equipment into the inventories of laboratories on a continuous basis. Thus, in addition to efficient sample management, the flexible structure is a vital feature of a reliable LIMS that will enable necessary integration with a wide range of laboratory instruments and applications.

Even with judicious use of modern sample tracking systems, the risk of sample error will always be present (Hofgärtner & Tait, 1999). It has been reported that 1.7 errors per 1000 investigations (excluding errors in test selection and interpretation) occur during genetic testing. The sample mix-up ratio has been found to be as high as 3% in some large-scale genotyping studies (Westra et al., 2011). Such errors could occur either before or after the sample arrives in the laboratory. To minimize sample error risk, laboratories can utilize additional methods. These methods may include requesting multiple patient identifiers in the order form and sample container (i.e., CAP requires at least two patient identifiers), employing automated systems, comparing genotype data with reported gender, and performing additional genotyping of a small number of variants such as single nucleotide polymorphisms (SNPs) and short tandem repeats (STRs) to confirm sample identity.

## DNA quality assessment

After nucleic acid extraction and purification, the quantity and quality of the sample needs to be evaluated. This is routinely performed using a combination of methods such as absorbance, fluorescence, and electrophoresis. Each method has its own strengths and limitations. Absorbance measurements permit detection of contaminants, such as carryover of the extraction reagents, but will not differentiate between nucleic acid types such as DNA or RNA due to their overlapping absorbance profile (Wilfinger et al., 1997). Fluorometric applications permit quantification of specific target nucleic acid species, including double-stranded versus single-stranded DNA, but accuracy depends on reliable standard curve input and this method is unable to assess the presence of any contaminating substances. Polymerase Chain Reaction (PCR) technologies and electrophoresis may be used to accurately measure low yield samples or assess fragment length and integrity, respectively, but both have similar purity assessment limitations as seen with fluorometric analysis (Robin et al., 2016).

The degree and sensitivity to which DNA quantity, quality, and integrity need evaluation depends on the downstream application. For example, quantitative PCR (qPCR) and targeted sequencing applications might only require routine quantification and contamination assessment as provided by absorbance measurements, with or without fluorometric assessment. In contrast, NGS, long-range PCR, and Third-Generation Sequencing (TGS) technologies require input of DNA fragments with a specific target size range, necessitating the additional determination of mean fragment length by electrophoresis in addition to quantitative and qualitative assessments.

## Testing platforms

Clinical genetics laboratories employ several genetic testing platforms, including qPCR (also known as real-time PCR), array-based systems, and sequencing-based methods (e.g., Sanger sequencing, NGS) to identify various genomic variants associated with drug response. Each testing method has advantages and disadvantages and the appropriate choices for each laboratory are based on individual needs. Although most pharmacogenomic variants are single nucleotide changes (i.e., SNPs), there are other types of variants such as small insertions or deletions and CNVs. Before clinical use, laboratories should determine the best methods to meet their needs and validate their selected methods to identify each variant included in the pharmacogenomic test panel. The following section provides a brief overview of common pharmacogenomic testing methods. Table 3.4 summarizes their advantages and disadvantages.

*Q-PCR* is a PCR-based method that enables monitoring the amplification of a targeted DNA sequence simultaneously (in real-time) with the PCR. Since the goal is identifying a specific sequence variant (e.g., SNP), sequence-specific DNA probes are needed. Drug metabolism genotyping assays designed for qPCR are commercially available and suitable for generating custom pharmacogenomic panels. In addition to specific SNP detection, qPCR can also be used to determine the number of copies present in pharmacogenomic genes with clinically relevant CNVs such as *CYP2D6*.

*Array-Based Platforms* are designed to detect up to several million SNPs, including tens of thousands of biomarkers relevant to drug response. In addition to determining variants at specific loci, array-based platforms enable detecting copy number changes and blocks with absence of heterozygosity (AOH) in the genome. On the other hand, this technology can only

**Table 3.4** Comparison of the advantages and limitations of different testing platforms.

| | Advantages | Disadvantages/Limitations |
|---|---|---|
| Q-PCR | - High sensitivity—High-throughput with large sample capacity—Short TAT—CNV analysis possible | - Multiplexing capacity is limited compared to next-generation sequencing |
| Array-based Methods | - High-throughput with large sample capacity—CNV analysis possible—enables AOH detection | - Interrogate variants only included in the array design—novel variant detection is not possible—Requires high quality DNA |
| MLPA | - Can detect single exon copy number change—needs 20 ng DNA—High throughput with short TAT—Requires simple laboratory equipment (thermocycler and capillary electrophoresis) | - Sensitive to DNA quality, concentration and extraction method—Sensitive to contamination—SNPs in the target region can cause false negative results |
| Sanger Sequencing | - Simple sample preparation process | - Low-throughput—not convenient for large sample size and multiple site genotyping |
| Targeted NGS | - Pre—designed single assay that analyzes—SNPs, InDel variations, and CNVs—fast TAT—Targeted testing for current treatment | - Genotyping limited to the targeted gene(s) |
| WES | - Higher coverage in targeted regions—CNV analysis possible (coding regions only) | - Provides data from coding regions of the human genome (1%–2%)—Long TAT |
| WGS | - Most comprehensive data—CNV analysis possible | - Higher cost—Long TAT—huge data size with novel variants of unknown effect |
| LRS | - High sequence range (>10 Kb)—useful for more accurate haplotyping and phasing | - Limited by the fragment sizes produced during the DNA extraction—Higher cost—Lower throughput—Higher error rate relative to SRS |

interrogate variants included in the original array design; thus, novel variant detection is impossible.

*Multiplex Ligation-Dependent Probe Amplification (MLPA)* is a highly sensitive multiplex PCR technique that is used for detection of CNVs with a range of sizes from a single exon to an entire chromosome. MLPA uses universal primer pairs to amplify fluorescently labeled unique probes of different lengths. Probes are designed in two pieces which hybridize adjacent to the target genomic sequence and require a ligation reaction in order to permit PCR amplification (e.g., "ligation-dependent"). Amplicons are fluorescently labeled and then separated by size and quantified by capillary electrophoresis.

Sequencing-based technologies include Sanger sequencing, targeted NGS (e.g., targeted gene panels), whole-exome sequencing (WES), and whole-genome sequencing (WGS). The latter two were also recently named exome sequencing and genome sequencing, respectively.

*Sanger Sequencing* was first introduced by Frederick Sanger in 1977 using the chain termination method and has been the standard gold method for many years (van Dijk et al., 2018). Sanger sequencing requires a simple sample preparation process, and can generate up to a 1 Kb long DNA product. However, it is low-throughput, cumbersome for evaluating a large number of samples, and not convenient for assaying multiple sites or for genome wide variant detection. Nevertheless, it can be useful for confirmation of debatable variants identified in qPCR or NGS data.

*Next-Generation Sequencing (NGS)* has been extensively used in the research and diagnosis of inherited diseases and has become an effective and prevalent method in precision medicine. NGS is massively high-throughput and can provide single-nucleotide resolution, but requires significant technical knowledge at a relatively higher cost. Targeted NGS, WES, and WGS are increasingly used in determining personal pharmacogenomic profiles in the clinical setting (Morganti et al., 2019). Several different sequencing by synthesis (SBS) technologies are available today for NGS, including pyrosequencing, dye sequencing, and ion semiconductor sequencing.

WGS does not employ target enrichment and thus enables analyses of both coding and noncoding regions of the genome thus provides the most comprehensive data on an individual genome. In a clinical setting, read-depth coverage of WGS is usually 30X (Mauger et al., 2020). This coverage level is also preferred when an unknown genetic etiology is investigated in an affected individual. WGS platforms with lower coverage, such as 5X or even lower than 1X (e.g., 0.5X, 0.4X), have been shown to produce surprisingly useful data when used in healthy individuals for precision medicine purposes (Homburger et al., 2019). Low coverage (also known as "low pass") WGS has also been shown to successfully detect structural chromosomal rearrangements (Chaubey et al., 2020).

Long-read sequencing, also known as third-generation sequencing (TGS), has several advantages over NGS technologies that use short reads (short-read sequencing, SRS). Examples of these include Nanopore (Oxford Nanopore) and single molecule real-time sequencing technology (SMRT) by PacBio. The NGS technologies mentioned above produce sequence reads of up to 600bp in length and have limited ability to evaluate repetitive regions and structural variants. On the other hand, long-read sequencing technologies (LRS) can yield products larger than 10 Kb in size and serve as high-throughput platforms for characterization of genomes through generation of highly contiguous assemblies (Pollard et al., 2018). Several prior studies proposed technical feasibility of using LRS for the analysis of trinucleotide repeat expansion and human disease-associated chromosomal structural variations (Chaisson et al., 2015; Liu et al., 2017; Loomis et al., 2013; Merker et al., 2018). LRS technology could be a useful tool for more accurate haplotyping and phasing. However, potential read length is largely limited by the fragment sizes produced during the DNA extraction process (Ardui et al., 2018; Petersen et al., 2019) and clinical use is limited by several factors including the higher cost, lower throughput, and higher error rate relative to SRS.

## Copy number assessment

In addition to SNPs, CNVs also play an important role in the level of activity of many enzymes important to PGx. A recent report identified novel exon-level deletions and duplications in 201 pharmacogenes (Santos et al., 2018). Several of these are Cytochrome P450 genes, which have been extensively studied due to their role in absorption, distribution, metabolism, and excretion (ADME) of many clinical drugs. The following section uses the *CYP2D6* gene as an example to provide a brief overview of the methods used for CNV assessment, the technologies used to correlate CNV changes to specific star alleles, as well as the strengths and limitations of these methods.

*CYP2D6* is involved in the metabolism of 20%–25% of prescribed drugs in the United States and Europe (Petrović et al., 2020; Saravanakumar et al.,

2019). Moreover, *CYP2D6* is a highly polymorphic enzyme with over 100 known SNPs and CNVs affecting enzymatic activity. The number of functional *CYP2D6* alleles may vary from zero to four or more with a reported CNV frequency in the US population of 13% (Beoris et al., 2016; Chan et al., 2019; Del Tredici, 2018). Fig. 3.4 illustrates some of the copy number and structural variations known to occur in the *CYP2D6* locus (Nofziger et al., 2020). Briefly, *CYP2D6* may have CNV events resulting in

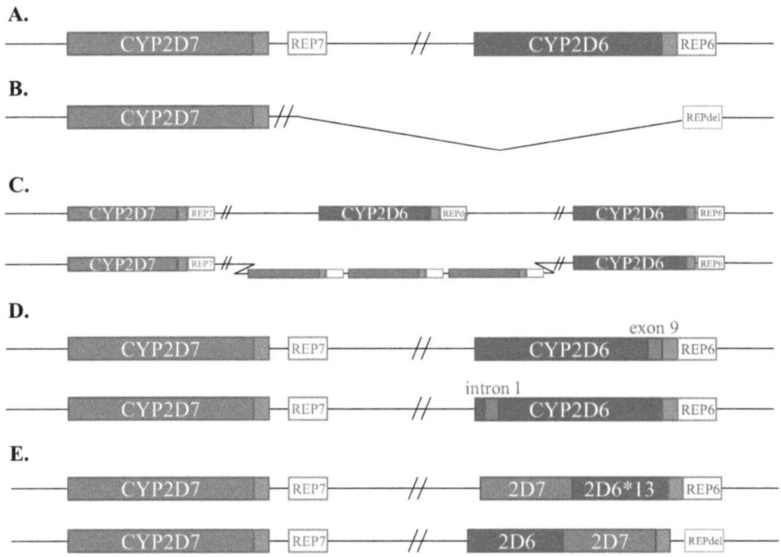


**Figure 3.4** *CYP2D6* Copy number and structural variations. The *CYP2D6* locus is commonly present as a tandem configuration of *CYP2D6* (blue (gray in print version) with the pseudogenes *CYP2D7* (orange (light gray in print version)) and *CYP2D8* (not shown) (A). The gray rectangles denote a shared 0.6 Kb long sequence at the 3'end of the genes that is followed by similar repeat sequences (REP). The repeat sequences can mediate *CYP2D6* whole gene deletion (B), known as a *5 allele, or duplication (C) events. In either case, the REP sequences generally deviate from the common state and are referred to as REPdel or REPdup (REPdp), respectively. CYP2D6 may also participate in complex rearrangements with *CYP2D7* (D & E). Small rearrangements can involve a single exon (e.g., exon 9 or 2) or intron, such as intron 1 (D). Larger conversion events can result in generation of *CYP2D7-2D6* or *CYP2D6-2D7* hybrid genes (E). *CYP2D7-2D6* hybrids can vary in the amount of *CYP2D6* that is replaced by *CYP2D7*, but all are generally nonfunctional and designated *13 alleles. The size of the *CYP2D6-2D7* hybrids can also vary, but at least one (*10.003) has some reported enzymatic activity. *From Nofziger, C., Turner, A. J., Sangkuhl, K., Whirl-Carrillo, M., Agúndez, J. A. G., Black, J. L., Dunnenberger, H. M., Ruano, G., Kennedy, M. A., Phillips, M. S., Hachad, H., Klein, T. E. & Gaedigk, A. (2020). PharmVar GeneFocus: CYP2D6.* Clinical Pharmacology and Therapeutics, 107*(1), 154–170. https://doi.org/10.1002/cpt.1643*

whole-gene deletion (known as *5 allele) or duplication (Del Tredici, 2018). Normal (e.g., *1, *2, and *35), low- (e.g., *17, *10, and *41), and no- (e.g., *3, *4, *6, and *36) function alleles have all been observed in gene duplication events (Sistonen et al., 2007; Del Tredici, 2018). Additionally, *CYP2D6* can participate in gene conversion events with a nearby pseudogene, *CYP2D7,* resulting in decreased or no function alleles (e.g., *13, *36, or *4). Therefore, whenever possible detected *CYP2D6* CNVs should be reported in the context of gene function. Indeed, a recent report emphasized the importance of *CYP2D6* in determining haplotype dosage along with copy number assessment for pharmacogenomic testing in psychiatry (Jarvis et al., 2019).

In a survey by CAP, nearly 10% of laboratories that test for *CYP2D6* indicated that they do not perform CNV testing (Pratt et al., 2021). Moreover, of those that do, the majority assess copy number at a single position and are thus susceptible to missing some complex rearrangement or fusion events with the *CYP2D7* pseudogene. This lack of consistency for evaluating CNVs in *CYP2D6* may result in erroneous allele determination, which adversely affects treatment decisions. Even if laboratories do assess copy number, several commonly used methods are not able to determine which specific allele is duplicated or multiplicated. This may result in an inability to generate an unambiguous phenotype assertion. For example, if the laboratory is unable to determine the total number of copies and which allele is affected, they may return a diplotype result of *1/*3 XN. Such a report indicates a status of intermediate, normal, or ultra-rapid metabolizer are all possible. When results are returned with such a wide range of possible phenotypes, determining how to utilize CDS poses multiple conundrums, including the inability to make clear dosing recommendations.

*Long-Range PCR* and *Southern blot* combined with *Sanger sequencing* were used early on to identify, characterize, and unambiguously correlate *CYP2D6* CNVs to observed haplotypes (Dorado et al., 2005; Gaedigk et al., 1991, 2007; Meijerman et al., 2007). The similarity of the pseudogene *CYP2D7* to *CYP2D6* impart an added burden to ensure sequencing of the intended target. Specifically, nested PCR of a full-gene length PCR product is required which, necessarily, delays the time to Sanger. Therefore, efforts have focused on identifying higher throughput methods to characterize this highly polymorphic locus in clinical diagnostic settings (Table 3.5), without the need to engage in such laborious, low throughput, and expensive stepwise assessments (Sheng et al., 2007).

**Table 3.5** Comparison of methods used for copy number variation assessment.

| Method | Sequencing platform | Strengths[1] | Weaknesses[1] |
|---|---|---|---|
| Long-Range PCR | – | Unambiguous haplotyping | Long TAT, laborious, painstaking |
| Southern blotting | – | Characterization of isolated DNA molecules | Radiation exposure, long TAT, laborious |
| qPCR and MLPA | – | Fast, reliable results at target site(s) | Bioinformatic haplotyping by inference |
| Pyrosequencing | – | Limit of detection (~5%) | Short reads (100–400bp); homopolymers; inability to differentiate between star alleles with the same SNP; requires haplotyping by inference |
| Sanger Sequencing | – | Moderate read lengths (800bp) | Limit of detection (~20%); requires haplotyping by inference |
| NGS | Illumina sequencers (various) and Thermo Fisher Ion Torrent | High sequence accuracy | Short reads necessitate bioinformatic haplotyping by inference |
| TGS | Pacific Biosciences and Oxford Nanopore | Long reads (≥10 kb) enables direct variant phasing and CN assessment | High capital cost, variable accuracy (92%–99%[2]), long TAT |

This table provides additional comparisons among the technologies for sequence listed in Table 5.4. (1) Technologies that use different chemistries or probes have been grouped due to their similar strengths and weaknesses; (2) Accuracy depends on platform and the application of computational corrections.

*Q-PCR*, often using TaqMan probes, has been used to evaluate *CYP2D6* CNVs and use of probes targeting exon 9 continues to a favored approach (Kisoi et al., 2020; Motamedi et al., 2012; Ramamoorthy et al., 2010; Stamer et al., 2002). However, using probes only targeting exon 9 risks missing *CYP2D7-2D6* hybrid alleles (Fig. 3.4), which result in nonfunctional hybrids (Black et al., 2012; Kramer et al., 2009). In order to more accurately phase *CYP2D6* duplication and *CYP2D7*-conversion events, semiquantitative allele ratios combined with CNV assessment at two distant sites within the gene (e.g., both intron 2 and exon 9) have been used (Table 3.5) (Chan et al., 2019; Del Tredici, 2018). This enables copy number (CN) assessment near the beginning and end of *CYP2D6*, which when combined with the relative ratio of variants detected by qPCR, allowing for more accurate phasing of the star alleles with variants within the copy number assessed region. For example, detecting a dosage of 1:2 for *1 to *3 and a copy number of 3 at both intron 2 and exon 9 will unambiguously correspond to a final activity score of 1 and an intermediate metabolizer phenotype for CYP2D6. However, this method still requires inference for variants that fall outside of the tested region (i.e., upstream of exon 2) or that are components of multiple star alleles.

Similarly, *MLPA* in conjunction with PCR-restriction fragment length polymorphism (RFLP) may aid evaluation of *CYP2D6* CNVs and SNPs while avoiding copy number misattribution that could result from co-targeting the *CYP2D7* pseudogene or inferring CN for a nonassessed region of the gene (Leandro-García et al., 2009). Still, efforts to phase variants and haplotype star alleles relative to CNV observed at discrete locations by qPCR or MLPA rely on inference based on the frequency of genotypes observed in *cis* or *trans*. Additionally, since the CNV assessment and SNP/INDEL assays are performed separately with qPCR, it may not be possible to assign the copy number to a specific star allele with absolute certainty. As a result, several approaches to characterize or sequence longer stretches of DNA in order to reliably correlate copy number to specific star alleles have been used.

*Pyrosequencing*, an SBS technology, has been used to both sequentially and simultaneously quantify and genotype *CYP2D6* (Langaee et al., 2015; Söderbäck et al., 2005). The benefit of using pyrosequencing to directly quantify the percentage of each allele present in a sample is the ability to accurately assign copy number to the specific variant. This information can then be used to infer the haplotype and corresponding phenotype. Observed limitations include an inability to differentiate *1/*1 from either *1/*5, *1/*10, or *1/*36 alleles, which share the same 100C > T SNP (Langaee et al., 2015).

*NGS*, using an SBS technology, has recently been used to characterize *CYP2D6* star alleles including CNVs. These technologies are often used to sequence the coding regions of multiple genes at one time (i.e., WES). Moreover, it is possible to bioinformatically assess the number of copies of every observed SNP by comparing the results to an internal reference. NGS technologies can be quite powerful. For example, WES data from a panel covering 340 pharmacogenes, combined with mRNA expression levels in human liver tissue samples, was recently used to identify over 600 deletions and duplications in 36 pharmacogenes (Tremmel et al., 2020).

Due to the short-read lengths of 100–600bp generated, the utility of NGS sequencing data for CNV assessment is largely dependent on the reliability of the bioinformatic pipelines used to analyze the data and assign calls into groupings that accurately reflect the patient's haplotype. Special algorithms have been developed for calling star alleles in highly polymorphic pharmacogenes, such as *CYP2D6*. These include Aldy, Astrolabe (formally Constellation), Cypiripi, PharmCAT, Stargazer, and VCF Annotator (Klein & Ritchie, 2018; Lee et al., 2019; Numanagić et al., 2015, 2018; Qiao et al., 2017; Twist et al., 2016). Aldy, Astrolabe, PharmCAT, and Stargazer are regularly maintained. Moreover, PharmCAT uses the Astrolabe allele calling output for *CYP2D6* clinical annotations, which are based on CPIC and DPWG guidelines (Caudle et al., 2020; Sangkuhl et al., 2020).

Each of these bioinformatic algorithms has its strengths and limitations. For example, a study showed Astrolabe was able to accurately predict haplotypes from paired-end WGS reads >100bp with an analytical sensitivity and specificity of 97% and 95%, respectively (Twist et al., 2016). Yet, it had difficulty aligning reads in exon 2 of *CYP2D6*, the *CYP2D6*/*CYP2D7* intergenic region, as well as repetitive regions located upstream or downstream of the gene. A recently described bioinformatics pipeline, Cyrius, had an impressive reported accuracy for *CYP2D6* haplotypes with structural variations of 94% compared to 87% for Aldy and 76% for Stargazer (Chen et al., 2021). Moreover, by comparing the results against a truth set it is claimed that the accuracy of structural variant calls by Cyrius improved to 100%. For these reasons it maybe resonable to use mutiple algorithms playing of their strengths for a particular call type (e.g. haplotype phasing versus structural variation).

*TGS* can be used to directly determine phasing of variants physically linked on the same strand of DNA and thereby overcome the inherent limitation of efforts to use bioinformatics to phase short-read NGS sequences. Examples of these TGS technologies include Nanopore (Oxford Nanopore) and SMRT (PacBio), both of which have been used to characterize *CYP2D6* (Ammar; Ardui et al., 2018; Petersen et al., 2019; Qiao et al., 2016). Although both Nanopore and SMRT employ SBS-type technologies, SMRT detects the colorimetric signal released upon incorporation of labeled nucleotides during DNA synthesis, while Nanopore measures the change in ionic current that occurs when single-stranded nucleic acids pass through biological nanopores. As a result, the SMRT reads are accurate, with no signal degradation over time, and read-length is dictated by dissociation of the polymerase from the template, resulting in a normal distribution of read lengths averaging 3 kb and up to 20 kb (Roberts et al., 2013). While Nanopore sequencing relies on robust bioinformatics algorithms generate a call rate accuracy of >99% for the ~10–30 Kb length reads generated on average, yet since polymerase dissociation is not as much of a limitation and reads as long as 2.3 Mb have been reported (Amarasinghe et al., 2020; Buermans et al., 2017).

Efforts must be made by clinical laboratories to carefully assess CNVs in pharmacogenes that are subject to complex rearrangements and duplications, such as *CYP2D6*, and clearly identify the limitations of their testing. There are various approaches available to phase variants (i.e. determine haplotypes) and for copy number assessment. Depending on the methods used, genotyping, phasing, and copy number assessment can be done using complimentary approaches, such as Sanger sequencing and qPCR, or can be evaluated simultaneously using the same approach, as with NGS or TGS.

Because of complexities such as these, it can be a significant undertaking for an institute to build and maintain a pipeline for internal phenotype calls. Historically, many institutes had a team of experts to develop and maintain a method to call phenotypes, but this siloed approach has led to variable interpretation and reporting of phenotypes, even when the same genotypes were observed (Abou Diwan et al., 2019). Standardization and the availability of professional guidelines have addressed this issue to some extent. More recently, some institutes have begun utilizing a vendor to translate the variant data provided by the laboratory (i.e., make the haplotype and phenotype calls). These vendors usually have a team to identify and apply any updated gene–drug interactions to the translation services provided to clinical laboratories, clinicians, and patients. Vendors either comprehensively perform both the interpretations and clinical support for an organization or may customize these services based on the specific needs of the organization.

## Conclusion

Starting a new laboratory or carving out a new section of a laboratory to support pharmacogenomic testing requires research into federal, state, and local requirements. Laboratories that are already accredited or newly accredited need to continually monitor accrediting entities for new rules, regulations, and accreditation checklist updates. Personnel requirements for high complexity labs must be considered due to the nonwaived status of pharmacogenomic testing. Understanding the difference in nomenclature between genetic variants and star alleles is the foundation upon which informed decisions regarding everything from choosing the best testing platforms to how the genotype results are translated and subsequently reported to clinicians with relevant CDS. It is also important to understand what needs to be tested in order to provide meaningful results back to patients and providers. Selected testing platforms and associated equipment should meet the laboratory's present needs, while keeping in mind future needs. Each test type must also be clinically validated according to published guidelines and regulations before being clinically available. Depending on the methods used, genotyping, phasing, and copy number assessment can be done using complimentary approaches or simultaneously using one approach. Every testing platform has its own particular strengths and weaknesses, including the need for a robust bioinformatics workflow, and these should be thoroughly vetted to understand the limitations of what can be reported. Although there are a variety of cutting-edge technologies, the calculus of which to choose should also include a thorough analysis of the equipment and reagent costs, and whether the laboratory will reasonably anticipate being able to support those costs with sufficient sample volumes and reimbursement. Institutes must also consider the advantages and disadvantages of building a program independently or utilizing vendors or outside resources. Finally, each institution needs to consider the approaches and expertise needed to tackle any foreseeable obstacles and opportunities when planning the content of their CDS.

## References

Abdullah-Koolmees, H., van Keulen, A. M., Nijenhuis, M., & Deneer, V. H. M. (2021). Pharmacogenetics guidelines: Overview and comparison of the DPWG, CPIC, CPNDS, and RNPGx guidelines. *Frontiers in Pharmacology, 11*. https://doi.org/10.3389/fphar.2020.595219

Abou Diwan, E., Zeitoun, R. I., Abou Haidar, L., Cascorbi, I., & Khoueiry Zgheib, N. (2019). Implementation and obstacles of pharmacogenetics in clinical practice: An international survey. *British Journal of Clinical Pharmacology, 85*(9), 2076–2088. https://doi.org/10.1111/bcp.13999

Amarasinghe, S. L., Su, S., Dong, X., Zappia, L., Ritchie, M. E., & Gouil, Q. (2020). Opportunities and challenges in long-read sequencing data analysis. *Genome Biology, 21*(1). https://doi.org/10.1186/s13059-020-1935-5

Ammar, R. (1000). Long read nanopore sequencing for detection of HLA and CYP2D6 variants and haplotypes.

Ardui, S., Ameur, A., Vermeesch, J. R., & Hestand, M. S. (2018). Single molecule real-time (SMRT) sequencing comes of age: Applications and utilities for medical diagnostics. *Nucleic Acids Research, 46*(5), 2159–2168. https://doi.org/10.1093/nar/gky066

Badaoui, S., Hopkins, A. M., Rodrigues, A. D., Miners, J. O., Sorich, M. J., & Rowland, A. (2021). Application of model informed precision dosing to address the impact of pregnancy stage and CYP2D6 phenotype on foetal morphine exposure. *AAPS Journal, 23*(1). https://doi.org/10.1208/s12248-020-00541-1

Beoris, M., Wilson, J. A., Garces, J. A., & Lukowiak, A. A. (2016). CYP2D6 copy number distribution in the US population. *Pharmacogenetics and Genomics, 26*(2), 96–99. https://doi.org/10.1097/FPC.0000000000000188

Black, J. L., Walker, D. L., O'Kane, D. J., & Harmandayan, M. (2012). Frequency of undetected CYP2D6 hybrid genes in clinical samples: Impact on phenotype prediction. *Drug Metabolism and Disposition, 40*(1), 111–119. https://doi.org/10.1124/dmd.111.040832

Brosen, K. (2015). Pharmacogenetics of drug oxidation via cytochrome P450 (CYP) in the populations of Denmark, Faroe Islands and Greenland. *Drug Metabolism and Personalized Therapy, 30*(3), 147–163. https://doi.org/10.1515/dmdi-2014-0029

Buermans, H. P. J., Vossen, R. H. A. M., Anvar, S. Y., Allard, W. G., Guchelaar, H. J., White, S. J., den Dunnen, J. T., Swen, J. J., & van der Straaten, T. (2017). Flexible and scalable full-length CYP2D6 long amplicon PacBio sequencing. *Human Mutation, 38*(3), 310–316. https://doi.org/10.1002/humu.23166

Cartabuke, R. S., Tobias, J. D., Taghon, T., & Rice, J. (2014). Current practices regarding codeine administration among pediatricians and pediatric subspecialists. *Clinical Pediatrics, 53*(1), 26–30. https://doi.org/10.1177/0009922813498151

Caudle, K. E., Sangkuhl, K., Whirl-Carrillo, M., Swen, J. J., Haidar, C. E., Klein, T. E., Gammal, R. S., Relling, M. V., Scott, S. A., Hertz, D. L., Guchelaar, H. J., & Gaedigk, A. (2020). Standardizing CYP2D6 genotype to phenotype translation: Consensus recommendations from the clinical pharmacogenetics implementation consortium and Dutch pharmacogenetics working group. *Clinical and Translational Science, 13*(1), 116–124. https://doi.org/10.1111/cts.12692

Chaisson, M. J. P., Huddleston, J., Dennis, M. Y., Sudmant, P. H., Malig, M., Hormozdiari, F., Antonacci, F., Surti, U., Sandstrom, R., Boitano, M., Landolin, J. M., Stamatoyannopoulos, J. A., Hunkapiller, M. W., Korlach, J., & Eichler, E. E. (2015). Resolving the complexity of the human genome using single-molecule sequencing. *Nature, 517*(7536), 608–611. https://doi.org/10.1038/nature13907

Chan, W., Li, M. S., Sundaram, S. K., Tomlinson, B., Cheung, P. Y., & Tzang, C. H. (2019). CYP2D6 allele frequencies, copy number variants, and tandems in the population of Hong Kong. *Journal of Clinical Laboratory Analysis, 33*(1), e22634. https://doi.org/10.1002/jcla.22634

Chaubey, A., Shenoy, S., Mathur, A., Ma, Z., Valencia, C. A., Reddy Nallamilli, B. R., Szekeres, E., Stansberry, L., Liu, R., & Hegde, M. R. (2020). Low-pass genome sequencing: Validation and diagnostic utility from 409 clinical cases of low-pass genome sequencing for the detection of copy number variants to replace constitutional microarray. *Journal of Molecular Diagnostics, 22*(6), 823–840. https://doi.org/10.1016/j.jmoldx.2020.03.008

Chen, Y., Nie, F., Xie, S. Q., Zheng, Y. F., Dai, Q., Bray, T., Wang, Y. X., Xing, J. F., Huang, Z. J., Wang, D. P., He, L. J., Luo, F., Wang, J. X., Liu, Y. Z., & Xiao, C. L. (2021). Efficient assembly of nanopore reads via highly accurate and intact error correction. *Nature Communications, 12*(1). https://doi.org/10.1038/s41467-020-20236-7

CLIA, & C.L. (2021). *Clinical laboratory improvement act (42 USC263a).*

Cucoranu, I. C. (2016). Laboratory information systems management and operations. *Clinics in Laboratory Medicine, 36*(1), 51–56. https://doi.org/10.1016/j.cll.2015.09.006

van Dijk, E. L., Jaszczyszyn, Y., Naquin, D., & Thermes, C. (2018). The third revolution in sequencing technology. *Trends in Genetics, 34*(9), 666–681. https://doi.org/10.1016/j.tig.2018.05.008

Del Tredici, A. L., et al. (2018). Frequency of CYP2D6 alleles including structural variants in the United States. *Frontiers in Pharmacology, 9*(305). https://doi.org/10.3389/fphar.2018.00305. 29674966.

Dorado, P., Cáceres, M. C., Pozo-Guisado, E., Wong, M.-L., Licinio, J., & Llerena, A. (2005). Development of a PCR-based strategy for CYP2D6 genotyping including gene multiplication of worldwide potential use. *BioTechniques, 39*(4S), S571–S574. https://doi.org/10.2144/000112044

Gaedigk, A., Blum, M., Gaedigk, R., Eichelbaum, M., & Meyer, U. A. (1991). Deletion of the entire cytochrome P450 CYP2D6 gene as a cause of impaired drug metabolism in poor metabolizers of the debrisoquine/sparteine polymorphism. *American Journal of Human Genetics, 48*(5), 943–950.

Gaedigk, A., Ndjountché, L., Divakaran, K., Dianne Bradford, L., Zineh, I., Oberlander, T. F., Brousseau, D. C., McCarver, D. G., Johnson, J. A., Alander, S. W., Riggs, K. W., & Leeder, J. S. (2007). Cytochrome P4502D6 (CYP2D6) gene locus heterogeneity: Characterization of gene duplication events. *Clinical Pharmacology and Therapeutics, 81*(2), 242–251. https://doi.org/10.1038/sj.clpt.6100033

Green, R. C., Roberts, J. S., Cupples, L. A., Relkin, N. R., Whitehouse, P. J., Brown, T., Eckert, S. L., Butson, M., Sadovnick, A. D., Quaid, K. A., Chen, C., Cook-Deegan, R., & Farrer, L. A. (2009). Disclosure of APOE genotype for risk of Alzheimer's disease. *New England Journal of Medicine, 361*(3), 245–254. https://doi.org/10.1056/NEJMoa0809578

Hein, D. W., Fakis, G., & Boukouvala, S. (2018). Functional expression of human arylamine N -acetyltransferase NAT1*10 and NAT1*11 alleles: A mini review. *Pharmacogenetics and Genomics, 28*(10), 238–244. https://doi.org/10.1097/FPC.0000000000000350

Hofgärtner, W. T., & Tait, J. F. (1999). Frequency of problems during clinical molecular-genetic testing. *American Journal of Clinical Pathology, 112*(1), 14–21. https://doi.org/10.1093/ajcp/112.1.14

Homburger, J. R., Neben, C. L., Mishne, G., Zhou, A. Y., Kathiresan, S., & Khera, A. V. (2019). Low coverage whole genome sequencing enables accurate assessment of common variants and calculation of genome-wide polygenic scores. *Genome Medicine, 11*(1). https://doi.org/10.1186/s13073-019-0682-2

Hu, X. X., Zhou, Q., Lan, T., Huang, X. X., Liang, B. Q., Dai, D. P., Cai, J. P., & Hu, G. X. (2016). Functional characterization of 22 novel CYP2D6 variants for the metabolism of Tamoxifen. *Journal of Pharmacy and Pharmacology, 68*(6), 819–825. https://doi.org/10.1111/jphp.12556

Jarvis, J. P., Peter, A. P., & Shaman, J. A. (2019). Consequences of CYP2D6 copy-number variation for pharmacogenomics in psychiatry. *Frontiers in Psychiatry, 10.* https://doi.org/10.3389/fpsyt.2019.00432

Kalman, L. V., Agúndez, J. A. G., Appell, M. L., Black, J. L., Bell, G. C., Boukouvala, S., Bruckner, C., Bruford, E., Caudle, K., Coulthard, S. A., Daly, A. K., Del Tredici, A. L., Den Dunnen, J. T., Drozda, K., Everts, R. E., Flockhart, D., Freimuth, R. R., Gaedigk, A., Hachad, H., … Zanger, U. M. (2016). Pharmacogenetic allele nomenclature: International workgroup recommendations for test result reporting. *Clinical Pharmacology and Therapeutics, 99*(2), 172–185. https://doi.org/10.1002/cpt.280

Kisoi, M., Imai, M., Yamamura, M., Sakaguchi, Y., Murata, S., Ichikawa, A., & Kinoshita, K. (2020). Unique genotyping protocol of CYP2D6 allele frequency using real time quantitative PCR from Japanese healthy women. *Biological and Pharmaceutical Bulletin, 43*(5), 904–907. https://doi.org/10.1248/BPB.B19-00512

Klein, T. E., & Ritchie, M. D. (2018). PharmCAT: A pharmacogenomics clinical annotation tool. *Clinical Pharmacology and Therapeutics, 104*(1), 19–22. https://doi.org/10.1002/cpt.928

Knych, H. K., Baden, R. W., Gretler, S. R., & McKemie, D. S. (2019). Characterization of the in vitro CYP450 mediated metabolism of the polymorphic CYP2D6 probe drug codeine in horses. *Biochemical Pharmacology, 168*, 184–192. https://doi.org/10.1016/j.bcp.2019.07.005

Kramer, W. E., Walker, D. L., O'Kane, D. J., Mrazek, D. A., Fisher, P. K., Dukek, B. A., Bruflat, J. K., & Black, J. L. (2009). CYP2D6: Novel genomic structures and alleles. *Pharmacogenetics and Genomics, 19*(10), 813–822. https://doi.org/10.1097/FPC.0b013e3283317b95

Langaee, T., Hamadeh, I., Chapman, A. B., Gums, J. G., Johnson, J. A., & Brusgaard, K. (2015). A novel simple method for determining CYP2D6 gene copy number and identifying allele(s) with duplication/multiplication. *PLOS ONE, 10*(1), e0113808. https://doi.org/10.1371/journal.pone.0113808

Leandro-García, L. J., Leskelä, S., Montero-Conde, C., Landa, I., López-Jimenez, E., Letón, R., Seeringer, A., Kirchheiner, J., Cascón, A., Robledo, M., & Rodríguez-Antona, C. (2009). Determination of CYP2D6 gene copy number by multiplex polymerase chain reaction analysis. *Analytical Biochemistry, 389*(1), 74–76. https://doi.org/10.1016/j.ab.2009.03.021

Lee, S.b., Wheeler, M. M., Patterson, K., McGee, S., Dalton, R., Woodahl, E. L., Gaedigk, A., Thummel, K. E., & Nickerson, D. A. (2019). Stargazer: A software tool for calling star alleles from next-generation sequencing data using CYP2D6 as a model. *Genetics in Medicine, 21*(2), 361–372. https://doi.org/10.1038/s41436-018-0054-0

Lemasters, G. K., Khurana Hershey, G. K., Sivaprasad, U., Martin, L. J., Pilipenko, V., Ericksen, M. B., Burkle, J. W., Lindsey, M. A., Bernstein, D. I., Lockey, J. E., Gareri, J., Lubetsky, A., Koren, G., & Biagini Myers, J. M. (2015). N-acetyltransferase 1 polymorphism increases cotinine levels in Caucasian children exposed to secondhand smoke: The CCAAPS birth cohort. *Pharmacogenomics Journal, 15*(2), 189–195. https://doi.org/10.1038/tpj.2014.44

Liu, Q., Zhang, P., Wang, D., Gu, W., & Wang, K. (2017). Interrogating the \\unsequenceable\\ genomic trinucleotide repeat disorders by long-read sequencing. *Genome Medicine, 9*(1). https://doi.org/10.1186/s13073-017-0456-7

Loomis, E. W., Eid, J. S., Peluso, P., Yin, J., Hickey, L., Rank, D., McCalmon, S., Hagerman, R. J., Tassone, F., & Hagerman, P. J. (2013). Sequencing the unsequenceable: Expanded CGG-repeat alleles of the fragile X gene. *Genome Research, 23*(1), 121–128. https://doi.org/10.1101/gr.141705.112

Lopes, G. S., Bielinski, S. J., Moyer, A. M., Black, J. L., Jacobson, D. J., Jiang, R., Larson, N. B., & Sauver, J. L. S. (2020). Sex differences in associations between CYP2D6 phenotypes and response to opioid analgesics. *Pharmacogenomics and Personalized Medicine, 13*, 71–79. https://doi.org/10.2147/PGPM.S239222

Madadi, P., Ross, C. J. D., Hayden, M. R., Carleton, B. C., Gaedigk, A., Leeder, J. S., & Koren, G. (2009). Pharmacogenetics of neonatal opioid toxicity following maternal use of codeine during breastfeeding: A case-control study. *Clinical Pharmacology and Therapeutics, 85*(1), 31–35. https://doi.org/10.1038/clpt.2008.157

Marques, S. C., & Ikediobi, O. N. (2010). The clinical application of UGT1A1 pharmacogenetic testing: Gene– environment interactions. *Human Genomics, 4*(4), 238–249. https://doi.org/10.1186/1479-7364-4-4-238

Mauger, F., Horgues, C., Pierre-Jean, M., Oussada, N., Mesrob, L., & Deleuze, J. F. (2020). Comparison of commercially available whole-genome sequencing kits for variant detection in circulating cell-free DNA. *Scientific Reports, 10*(1). https://doi.org/10.1038/s41598-020-63102-8

Meijerman, I., Sanderson, L. M., Smits, P. H. M., Beijnen, J. H., & Schellens, J. H. M. (2007). Pharmacogenetic screening of the gene deletion and duplications of CYP2D6. *Drug Metabolism Reviews, 39*(1), 45–60. https://doi.org/10.1080/03602530600952206

Merker, J. D., Wenger, A. M., Sneddon, T., Grove, M., Zappala, Z., Fresard, L., Waggott, D., Utiramerur, S., Hou, Y., Smith, K. S., Montgomery, S. B., Wheeler, M., Buchan, J. G., Lambert, C. C., Eng, K. S., Hickey, L., Korlach, J., Ford, J., & Ashley, E. A. (2018). Long-read genome sequencing identifies causal structural variation in a Mendelian disease. *Genetics in Medicine, 20*(1), 159–163. https://doi.org/10.1038/gim.2017.86

Morganti, S., Tarantino, P., Ferraro, E., D'Amico, P., Duso, B. A., & Curigliano, G. (2019). Next generation sequencing (NGS): A revolutionary technology in pharmacogenomics and personalized medicine in cancer. *Advances in Experimental Medicine and Biology, 1168*, 9–30. https://doi.org/10.1007/978-3-030-24100-1_2. Springer.

Motamedi, S., Majidzadeh, K., Mazaheri, M., Anbiaie, R., Mortazavizadeh, S. M. R., & Esmaeili, R. (2012). Tamoxifen resistance and CYP2D6 copy numbers in breast cancer patients. *Asian Pacific Journal of Cancer Prevention, 13*(12), 6101–6104. https://doi.org/10.7314/APJCP.2012.13.12.6101

Neafsey, P., Ginsberg, G., Hattis, D., & Sonawane, B. (2009). Genetic polymorphism in cytochrome P450 2D6 (CYP2D6): Population distribution of CYP2D6 activity. *Journal of Toxicology and Environmental Health - B: Critical Reviews, 12*(5–6), 334–361. https://doi.org/10.1080/10937400903158342

Nofziger, C., & Paulmichl, M. (2018). Accurately genotyping CYP2D6: Not for the faint of heart. *Pharmacogenomics, 19*(13), 999–1002. https://doi.org/10.2217/pgs-2018-0105

Nofziger, C., Turner, A. J., Sangkuhl, K., Whirl-Carrillo, M., Agúndez, J. A. G., Black, J. L., Dunnenberger, H. M., Ruano, G., Kennedy, M. A., Phillips, M. S., Hachad, H., Klein, T. E., & Gaedigk, A. (2020). PharmVar GeneFocus: CYP2D6. *Clinical Pharmacology and Therapeutics, 107*(1), 154–170. https://doi.org/10.1002/cpt.1643

Numanagić, I., Malikić, S., Ford, M., Qin, X., Toji, L., Radovich, M., Skaar, T. C., Pratt, V. M., Berger, B., Scherer, S., & Sahinalp, S. C. (2018). Allelic decomposition and exact genotyping of highly polymorphic and structurally variant genes. *Nature Communications, 9*(1). https://doi.org/10.1038/s41467-018-03273-1

Numanagić, I., Malikić, S., Pratt, V. M., Skaar, T. C., Flockhart, D. A., & Sahinalp, S. C. (2015). Cypiripi: Exact genotyping of CYP2D6 using high-throughput sequencing data. *Bioinformatics, 31*(12), i27–i34. https://doi.org/10.1093/bioinformatics/btv232. Oxford University Press.

Petersen, L. M., Martin, I. W., Moschetti, W. E., Kershaw, C. M., Tsongalis, G. J., & Kraft, C. S. (2019). Third-generation sequencing in the clinical laboratory: Exploring the advantages and challenges of nanopore sequencing. *Journal of Clinical Microbiology, 58*(1). https://doi.org/10.1128/jcm.01315-19

Petrović, J., Pešić, V., & Lauschke, V. M. (2020). Frequencies of clinically important CYP2C19 and CYP2D6 alleles are graded across Europe. *European Journal of Human Genetics, 28*(1), 88–94. https://doi.org/10.1038/s41431-019-0480-8

Pollard, M. O., Gurdasani, D., Mentzer, A. J., Porter, T., & Sandhu, M. S. (2018). Long reads: Their purpose and place. *Human Molecular Genetics, 27*(R2), R234–R241. https://doi.org/10.1093/hmg/ddy177

Pratt, V. M., Cavallari, L. H., Del Tredici, A. L., Gaedigk, A., Hachad, H., Ji, Y., Kalman, L. V., Ly, R. C., Moyer, A. M., Scott, S. A., van Schaik, R. H. N., Whirl-Carrillo, M., & Weck, K. E. (2021). Recommendations for clinical CYP2D6 genotyping allele selection: A Joint consensus recommendation of the association for molecular pathology, College of American Pathologists, Dutch pharmacogenetics working group of the royal Dutch pharmacists association, and the European society for pharmacogenomics and personalized therapy. *Journal of Molecular Diagnostics, 23*(9), 1047–1064. https://doi.org/10.1016/j.jmoldx.2021.05.013

Pratt, V. M., Del Tredici, A. L., Hachad, H., Ji, Y., Kalman, L. V., Scott, S. A., & Weck, K. E. (2018). Recommendations for clinical CYP2C19 genotyping allele selection: A report of the association for molecular pathology. *Journal of Molecular Diagnostics, 20*(3), 269–276. https://doi.org/10.1016/j.jmoldx.2018.01.011

Qiao, W., Wang, J., Pullman, B. S., Chen, R., Yang, Y., & Scott, S. A. (2017). The CYP2D6 VCF translator. *Pharmacogenomics Journal, 17*(4), 301–303. https://doi.org/10.1038/tpj.2016.14

Qiao, W., Yang, Y., Sebra, R., Mendiratta, G., Gaedigk, A., Desnick, R. J., & Scott, S. A. (2016). Long-read single molecule real-time full gene sequencing of cytochrome P450-2D6. *Human Mutation, 37*(3), 315–323. https://doi.org/10.1002/humu.22936

Quality management for molecular genetic testing; approved guideline. CLSI document MM20-A (ISBN 1-56238-859-2). (n.d.).

Ramamoorthy, A., Flockhart, D. A., Hosono, N., Kubo, M., Nakamura, Y., & Skaar, T. C. (2010). Differential quantification of CYP2D6 gene copy number by four different quantitative real-time PCR assays. *Pharmacogenetics and Genomics, 20*(7), 451–454. https://doi.org/10.1097/FPC.0b013e32833a1083

Ramsey, L. B., Ong, H. H., Schildcrout, J. S., Shi, Y., Tang, L. A., Hicks, J. K., El Rouby, N., Cavallari, L. H., Tuteja, S., Aquilante, C. L., Beitelshees, A. L., Lemkin, D. L., Blake, K. V., Williams, H., Cimino, J. J., Davis, B. H., Limdi, N. A., Empey, P. E., Horvat, C. M., Kao, D. P., Lipori, G. P., Rosenman, M. B., Skaar, T. C., Teal, E., Winterstein, A. G., Owusu Obeng, A., Salyakina, D., Gupta, A., Gruber, J., McCafferty-Fernandez, J., Bishop, J. R., Rivers, Z., Benner, A., Tamraz, B., Long-Boyle, J., Peterson, J. F., & Van Driest, S. L. (2020). IGNITE Pharmacogenetics Working Group. Prescribing prevalence of medications with potential genotype-guided dosing in pediatric patients. *Journal of the American Medical Association Network Open, 3*(12), Article e2029411. https://doi.org/10.1001/jamanetworkopen.2020.29411

Roberts, R. J., Carneiro, M. O., & Schatz, M. C. (2013). The advantages of SMRT sequencing. *Genome Biology, 14*(6). https://doi.org/10.1186/gb-2013-14-6-405

Robin, J. D., Ludlow, A. T., La Ranger, R., Wright, W. E., & Shay, J. W. (2016). Comparison of DNA quantification methods for next generation sequencing. *Scientific Reports, 6*. https://doi.org/10.1038/srep24067

Sangkuhl, K., Whirl-Carrillo, M., Whaley, R. M., Woon, M., Lavertu, A., Altman, R. B., Carter, L., Verma, A., Ritchie, M. D., & Klein, T. E. (2020). Pharmacogenomics clinical annotation tool (PharmCAT). *Clinical Pharmacology and Therapeutics, 107*(1), 203–210. https://doi.org/10.1002/cpt.1568

Santos, M., Niemi, M., Hiratsuka, M., Kumondai, M., Ingelman-Sundberg, M., Lauschke, V. M., & Rodríguez-Antona, C. (2018). Novel copy-number variations in pharmacogenes contribute to interindividual differences in drug pharmacokinetics. *Genetics in Medicine, 20*(6), 622–629. https://doi.org/10.1038/gim.2017.156

Saravanakumar, A., Sadighi, A., Ryu, R., & Akhlaghi, F. (2019). Physicochemical properties, biotransformation, and transport pathways of established and newly approved medications: A systematic review of the top 200 most prescribed drugs vs. the FDA-approved

drugs between 2005 and 2016. *Clinical Pharmacokinetics, 58*(10), 1281–1294. https://doi.org/10.1007/s40262-019-00750-8
Sheng, H. H., Zeng, A. P., Zhu, W. X., Zhu, R. F., Li, H. M., Zhu, Z. D., Qin, Y., Jin, W., Liu, Y., Du, Y. L., Sun, J., & Xiao, H. S. (2007). Allelic distributions of CYP2D6 gene copy number variation in the Eastern Han Chinese population. *Acta Pharmacologica Sinica, 28*(2), 279–286. https://doi.org/10.1111/j.1745-7254.2007.00479.x
Sistonen, J., Sajantila, A., Lao, O., Corander, J., Barbujani, G., & Fuselli, S. (2007). CYP2D6 worldwide genetic variation shows high frequency of altered activity variants and no continental structure. *Pharmacogenetics and Genomics, 17*(2), 93–101. https://doi.org/10.1097/01.fpc.0000239974.69464.f2
Söderbäck, E., Zackrisson, A.-L., Lindblom, B., & Alderborn, A. (2005). Determination of CYP2D6 gene copy number by pyrosequencing. *Clinical Chemistry, 51*(3), 522–531.
Stamer, U. M., Bayerer, B., Wolf, S., Hoeft, A., & Stüber, F. (2002). Rapid and reliable method for cytochrome P450 2D6 genotyping. *Clinical Chemistry, 48*(9), 1412–1417. https://doi.org/10.1093/clinchem/48.9.1412
Stocco, G., Cuzzoni, E., De Iudicibus, S., Favretto, D., Malusà, N., Martelossi, S., Pozzi, E., Lionetti, P., Ventura, A., & Decorti, G. (2015). Thiopurine metabolites variations during co-treatment with aminosalicylates for inflammatory bowel disease: Effect of N-acetyl transferase polymorphisms. *World Journal of Gastroenterology, 21*(12), 3571–3578. https://doi.org/10.3748/wjg.v21.i12.3571
Tremmel, R., Klein, K., Battke, F., Fehr, S., Winter, S., Scheurenbrand, T., Schaeffeler, E., Biskup, S., Schwab, M., & Zanger, U. M. (2020). Copy number variation profiling in pharmacogenes using panel-based exome resequencing and correlation to human liver expression. *Human Genetics, 139*(2), 137–149. https://doi.org/10.1007/s00439-019-02093-7
Twist, G. P., Gaedigk, A., Miller, N. A., Farrow, E. G., Willig, L. K., Dinwiddie, D. L., Petrikin, J. E., Soden, S. E., Herd, S., Gibson, M., Cakici, J. A., Riffel, A. K., Leeder, J. S., Dinakarpandian, D., & Kingsmore, S. F. (2016). Constellation: A tool for rapid, automated phenotype assignment of a highly polymorphic pharmacogene, CYP2D6, from whole-genome sequences. *Npj Genomic Medicine, 1*(1). https://doi.org/10.1038/npjgenmed.2015.7
Virbalas, J., Morrow, B. E., Reynolds, D., Bent, J. P., & Ow, T. J. (2019). The prevalence of ultrarapid metabolizers of codeine in a diverse urban population. *Otolaryngology - Head and Neck Surgery (United States), 160*(3), 420–425. https://doi.org/10.1177/0194599818804780
Westra, H. J., Jansen, R. C., Fehrmann, R. S. N., te Meerman, G. J., van Heel, D., Wijmenga, C., & Franke, L. (2011). MixupMapper: Correcting sample mix-ups in genome-wide datasets increases power to detect small genetic effects. *Bioinformatics, 27*(15), 2104–2111. https://doi.org/10.1093/bioinformatics/btr323
Wilfinger, W. W., Mackey, K., & Chomczynski, P. (1997). Effect of pH and ionic strength on the spectrophotometric assessment of nucleic acid purity. *BioTechniques, 22*(3), 474–481. https://doi.org/10.2144/97223st01
Williams, D. G., Patel, A., & Howard, R. F. (2002). Pharmacogenetics of codeine metabolism in an urban population of children and its implications for analgesic reliability. *British Journal of Anaesthesia, 89*(6), 839–845. https://doi.org/10.1093/bja/aef284
Yang, Y., Botton, M. R., Scott, E. R., & Scott, S. A. (2017). Sequencing the CYP2D6 gene: From variant allele discovery to clinical pharmacogenetic testing. *Pharmacogenomics, 18*(7), 673–685. https://doi.org/10.2217/pgs-2017-0033
Yue, Q. Y., Svensson, J. O., Säwe, J., & Bertilsson, L. (1995). Codeine metabolism in three oriental populations: A pilot study in Chinese, Japanese and Koreans. *Pharmacogenetics, 5*(3), 173–177. https://doi.org/10.1097/00008571-199506000-00006

CHAPTER FOUR

# Advancing equity in the promise of pharmacogenomics

**Alison E. Fohner[1], Shayna R. Killam[2], Kyle G. Volk[3], Erica L. Woodahl[2]**
[1]Department of Epidemiology and Institute of Public Health Genetics, University of Washington, Seattle, WA, United States
[2]Department of Biomedical and Pharmaceutical Sciences and Skaggs Institute for Health Innovation, University of Montana, Missoula, MT, United States
[3]Department of History, University of Montana, Missoula, MT, United States

## Contents



## Introduction

Pharmacogenomics is at a crossroads. There is great potential for pharmacogenomics to improve patient outcomes and to reduce inequities in medication prescribing, yet major challenges lie ahead for this promise to be broadly realized. Existing knowledge of global variation in genes known to affect drug response is woefully incomplete because the research community has largely focused recruitment efforts for pharmacogenomic studies on the same types of populations—people who are overwhelmingly wealthy, European-descended, and with consistent access to cutting-edge healthcare (Fohner et al., 2019). Without thoughtful consideration of the limitations of

*Clinical Decision Support for Pharmacogenomic Precision Medicine*
ISBN: 978-0-12-824453-1
https://doi.org/10.1016/B978-0-12-824453-1.00011-7

existing pharmacogenomic data and purposeful efforts to ameliorate them, implementing pharmacogenomic-based clinical decision support (CDS) has the potential to increase healthcare disparities. Ensuring pharmacogenomics improves health equity for all—instead of exacerbating inequalities—requires an ethical commitment to health justice, the inclusion of diverse research participants, and a change in public policy and insurance practices to ensure that pharmacogenomic testing is accessible to all patients regardless of socioeconomic status. This chapter highlights 1) the ways pharmacogenomic CDS can advance healthcare equity; 2) the ongoing demands for improving the quality of pharmacogenomic CDS tools and the data underlying clinical recommendations; and 3) the strategies available for users of pharmacogenomic CDS to reduce inequities in the quality of and access to these tools in their own healthcare settings. Several groups are working to address the historical biases in generalizability of pharmacogenomic data. Boxes throughout the chapter describe some of these efforts to reduce inequities in pharmacogenomic CDS and related disparities, including those working to increase participation of underrepresented populations and diverse practice settings in pharmacogenomics research.

## Pharmacogenomic CDS to advance healthcare equity

Pharmacogenomics offers a solution to several challenges in healthcare delivery that have historically led to inequities in care. It promises to personalize and target medication selection and dosing based on individual patient genetics, rather than on population averages or empirical observation. By facilitating decision-making around the use of pharmacogenomic tests, CDS tools are designed to help providers optimize treatment for individuals in less time and in fewer medical visits. CDS can make genetic information accessible to providers through the electronic health record (EHR) and guide them on how to use pharmacogenomic tests effectively, even if providers lack expertise in genomics or pharmacogenomics. Patients without access to pharmacogenomics may require more time and healthcare resources to identify an effective medication treatment plan or may never receive optimal drug therapy (Abul-Husn & Kenny, 2019; Wake et al., 2021). This section describes the potential of CDS to reduce healthcare disparities and improve health outcomes for all people by enabling treatment based on individual genetics and by eliminating biases in access to healthcare and in treatment guidelines.

## Improved access to advanced treatments

Pharmacogenomic CDS can reduce the gap in disparities that results from worse healthcare access. Historically, patients who live in rural areas that are geographically remote from larger health systems, are of lower socioeconomic status, are living with disabilities, or lack proper transportation to healthcare facilities have reduced access to healthcare and thus experience worse health outcomes. Pharmacogenomic CDS can mitigate the consequences of this limited access by lessening the time needed to identify an appropriate treatment regimen. This may be particularly beneficial for underserved and at-risk patients who may be less able to return for a follow-up visit. Benefits of achieving optimized medication regimens quickly include reduced hospitalization rates, improved therapeutic outcomes, improved patient adherence, and increased savings for patients, insurance companies, and healthcare organizations. Thus, pharmacogenomics may address the healthcare disparities often experienced by underserved patients, who are less able to meet the high time and cost demands of achieving therapy benefit using non-pharmacogenomic approaches.

As an example, the majority of communities in the state of Montana in the United States are classified as health professional shortage areas, with limited access to primary care and especially to specialty care. Two-thirds of the population in Montana live in rural areas and patients living hundreds of miles away from their healthcare providers may only be seen and assessed 2 to 3 times per year. Follow-up can be more difficult for these patients contributing to poorer clinical outcomes. For instance, in psychiatry, optimizing medications often requires repeat office visits and takes weeks to months to achieve maximum therapeutic benefit, which may be possible in urban centers but prohibitive in geographically remote communities. Many of the drugs used to treat psychiatric disorders exhibit large interindividual differences in drug disposition and response, much of which can be attributed to genetic variation (Brown et al., 2019; Hicks et al., 2015, 2017). Implementation of pharmacogenomic CDS in these settings could be especially valuable for achieving effective treatment for these remote patients.

In addition to improving patient access to healthcare, pharmacogenomic CDS may overcome limitations of healthcare delivery capacity. For example, many providers in smaller clinics away from major medical centers have limited training in genomics. As a result, even if providers have patient-specific genomic data, they may lack the expertise or resources needed to interpret and apply the information to patient care. CDS may enhance provider utilization of pharmacogenomics testing by providing a framework readily available through the EHR platform that assists in interpreting and applying pharmacogenomic results. Thus, more patients in a diverse array of practice settings may receive optimized therapies and achieve better health outcomes as pharmacogenomic CDS democratizes the use of pharmacogenomic information.

## Departure from race-based medicine

Advancements in genomics research enable clinical decisions to be specific to individual patient profiles, rather than be derived from a patient's race or ethnicity. Race-based guidelines can further exacerbate disparities by recommending alternative care for a subset of patients based on inadequate data and by limiting utility of the guidelines for populations that are underrepresented in research. By identifying and characterizing the full spectrum of therapeutically relevant genetic variants in an individual, pharmacogenomics-based dose adjustments can be more accurate than group-based guidelines. Pharmacogenomic CDS can reduce health disparities by recommending the appropriate drug dose and choice in all patients, regardless of ancestry.

For several medications, such as warfarin, average dosing requirements differ based on race or ancestry, even after accounting for common clinical factors such as weight, sex, age, and diet. On average, patients of Asian descent require a weekly dose of approximately 25 mg of warfarin, whereas patients of European descent require an estimated 37 mg per week and patients of African descent require approximately 45 mg each week (Dang et al., 2005). This variation in population dosing requirements can be explained in part by differences in population frequencies of pharmacogenomic variants known to affect warfarin response. Inclusion of genetic variants in a warfarin dosing algorithm accounts for 48%—74% of dose variation compared to only 20% accounted for by clinical factors alone (Gladding et al., 2010). More specifically, the −1639A allele (rs9923231) in the vitamin K epoxide reductase (*VKORC1*) gene, which encodes the drug target of warfarin, is associated with lower warfarin doses: patients homozygous for the A allele require a 100% lower dose than patients with no copy of the allele (Yang et al., 2010). The frequency of the *VKORC* −1639A allele is 41% in European populations and 88% in East Asian populations, and largely explains the difference in population average weekly doses of warfarin (Johnson et al., 2011, 2017; Lam & Cheung, 2012). Average doses in a population, however, do not provide information on the dose needed for an individual. For instance, some patients of Asian descent who carry the *VKORC1* −1639G allele require a higher dose of warfarin than would be expected based on population averages alone.

The frequency of the human leukocyte antigen (*HLA*), *HLA-A* and *HLA-B*, alleles associated with severe drug-induced cutaneous adverse reactions to abacavir, allopurinol, carbamazepine, oxcarbazepine, and phenytoin, including Stevens–Johnson syndrome and toxic epidermal necrolysis also varies according to race or ancestry (Caudle et al., 2014; Hershfield et al., 2013; Karnes et al., 2021; Leckband et al., 2013; Martin et al., 2014; Phillips et al., 2018; Saito et al., 2016). The frequencies of *HLA-A* and *HLA-B* alleles vary widely across the globe, with some Asian populations having a particularly high prevalence of many of these alleles, and thus, higher incidence of severe adverse reactions to these medications. Instead of avoiding these medications altogether in participants of Asian descent, because of the increased risk of adverse reactions, all patients could be genotyped with alternative medications used in patients who have at least one of the risk *HLA-A* or *HLA-B* alleles. With genotype information in the EHR, these medications could be used safely in patients who do not possess risk alleles, including those of Asian descent.

While the frequencies of genetic variants varies depending on ancestry, individual patients may or may not have those variants that are most common in the population most closely aligned with their ancestry. Furthermore, designating racial groups for patients is notoriously problematic and leads to misclassification (Roberts, 2011; Saini, 2019). Genomic profiles may not reflect the ancestry with which a patient most readily identifies, their individual admixture may make selection of one ancestry difficult, patients may be unfamiliar with their family history or ancestry, and healthcare providers may assume ancestry incorrectly. Thus, by testing individual pharmacogenomics profiles, rather than relying on population averages of a racial group and probable genotypes, medication dosing can be more accurate for all patients and will avoid broad and potentially incorrect decisions about drug selection based on race.

## Potential for pharmacogenomics to increase healthcare disparities

As with other emerging healthcare tools, the potential benefit of pharmacogenomics is limited by research paradigms and healthcare infrastructure, with current efforts primarily benefitting a small subset of people:

patients of European ancestry, those receiving care in large health systems, and those affluent enough to afford specialized care. In this way, pharmacogenomic CDS tools also have the potential to exacerbate healthcare disparities if selected patients are more likely to have access to them. With disparities of wealth and global socioeconomic stratification on the rise, the need for researchers and healthcare providers to be aware of this potential and address health equity through the development and dissemination of mature pharmacogenomic CDS systems is even more pressing (Fohner et al., 2019; Luczak et al., 2021; Magavern et al., 2022). In this section, we describe ways current pharmacogenomic research and implementation may increase disparities in healthcare and health outcomes.

## Limited population diversity in pharmacogenomic research

Engaging diverse populations in genomic research is crucial for obtaining comprehensive knowledge of variation in the human genome. Yet genomic research, including pharmacogenomics, has historically been biased toward inclusion of people of European ancestry. While global efforts have aimed to increase the diversity of genomic research participants—with the largest increases in populations of Asian ancestry—participants with African, Hispanic, Indigenous, Middle Eastern, and Oceanian ancestry remain troublingly underrepresented (Martin et al., 2017; Popejoy & Fullerton, 2016). This bias is problematic because the prevalence and frequency of variants in pharmacogenes genes is highly variable across populations (Huddart et al., 2019; McInnes et al., 2020). As a result, variants frequently found in European populations are relatively well characterized, while knowledge of common and important variants in non-European populations is more limited.

Several international groups, including the Clinical Pharmacogenetics Implementation Consortium (CPIC) and the Dutch Pharmacogenomics Working Group (DPWG), synthesize existing pharmacogenomics research and author evidence-based guidelines for genotype-guided dosing (Whirl-Carrillo et al., 2021). In many cases, guidance from these groups provides the evidentiary foundation for existing pharmacogenomic CDS tools. These groups rely on published data to make dosing recommendations for specific genetic variants, but variants that are underrepresented in the literature may not be sufficiently characterized to reliably predict how they affect drug response or toxicity. As a result, variants in underrepresented populations may not receive a recommendation in the guidelines, which further contributes to the persistence of this bias into clinical implementation and pragmatic studies of clinical utility.

Warfarin provides another example of the consequences of these biases (Kaye et al., 2017; Perera et al., 2014). CPIC dosing recommendations for warfarin recommend dose adjustments based in part on genetic variation in the cytochrome P450 2C9 (*CYP2C9*) gene, as well as *VKORC1* (Johnson et al., 2011, 2017). CPIC guidelines published in 2011 included strong recommendations for the *CYP2C9*2* and **3* alleles, which are most common in populations of European ancestry (Johnson et al., 2011). The most common alleles identified in populations of African ancestry, *CYP2C9*5*, **6*, **8*, and **11* alleles, were not included in the recommendations due to insufficient evidence. As a result, patients with these variants did not receive the same clinical benefit from pharmacogenomic testing. While the 2017 update to the guidelines included dosing recommendations for these variants based on additional data, these recommendations were still classified as moderate or optional (Johnson et al., 2017). Increased research targeted to include populations of African ancestry, where these variants are more common, led to more comprehensive guidelines. Warfarin is one of the best characterized medications with pharmacogenomic recommendations, and yet even it is plagued by such biases.

The antiplatelet agent clopidogrel provides another example of the potential of inadequately diverse pharmacogenomic data to exacerbate poor health outcomes. The cytochrome P450 2C19 (CYP2C19) enzyme activates clopidogrel to its active metabolites and *CYP2C19* genetic variation influences enzyme function, with frequencies varying widely across populations. CPIC guidelines recommend that patients with no function *CYP2C19* alleles (e.g., *CYP2C19*2* and **3*) receive an alternative antiplatelet therapy because of the increased risk of cardiovascular events due to decreased bioactivation of clopidogrel (Scott et al., 2011, 2013). In 2021, the state of Hawaii was granted a financial settlement of $834 million against Sanofi and Bristol Myers Squibb, the manufacturers of clopidogrel, for deceptive marketing practices that failed to warn patients of the health risks associated with clopidogrel (Bellon, 2021). The lawsuit argued that the companies ignored genetic data suggesting an increased risk for Hawaiian residents and continued to market clopidogrel as effective, resulting in subsequent harm to patients. The state of Hawaii prevailed in their argument, in part, because specific pharmacogenomic data on residents in Hawaii do not exist.

In Hawaii, the three largest groups by ancestry are Asian, European, and Native Hawaiian or other Pacific Islander. Pharmacogenomic frequency data include an Oceanian ancestry group that is defined as "pre-colonial populations of the Pacific Islands, including Hawaii, Australia, and Papua New Guinea" (Huddart et al., 2019). Allele frequencies of *CYP2C19*2* are the highest in Oceanian populations (61%), followed by East (28%) and Central/South Asian (27%), and comparatively low in European (15%) (Botton et al., 2021; Scott et al., 2013). Similarly, *CYP2C19**3 frequencies are highest in Oceanian populations (15%), followed by East (7%) and Central/South Asian (2%), with very low frequencies in European (0.2%). Taken together, one could hypothesize from these frequencies that a large proportion of Hawaiian residents will carry *CYP2C19* alleles with no function, leading to therapeutic failure of clopidogrel and an increased risk of cardiovascular events. The lawsuit contended that pharmacogenomic-guided clopidogrel prescribing may have been particularly important in Hawaiian patients given the high frequency of no function alleles in Oceanian and Asian populations, but prescribers were not aware of the potential risks associated with clopidogrel in their patients. It should be noted, however, that the populations in Hawaii are highly admixed and at present there are no published pharmacogenomics studies specific to Native Hawaiians or non-Native Hawaiians. Uncharacterized variants may also be common in these populations. In this case, diverse research participation from Hawaiian participants may have reduced the harm described in the lawsuit by providing actionable and tailored pharmacogenomic recommendations to adjust standard dosing protocols in this patient population.

It is essential for researchers to include diverse populations in research, for providers to be aware of the populations excluded from research, and for guidelines to be transparent regarding the population demographics from which pharmacogenomic CDS tools are generated. The majority of pharmacogenomic genotyping platforms target only a subset of variants with known function—largely those identified to be predictive in populations of European ancestry—and therefore, may miss relevant variation in understudied populations. As a result, people with variants that are not included in these platforms—of any ancestry—could receive incorrect or incomplete genotype information, leading to inaccurate phenotype predictions and dosing recommendations. Providers must be aware that CDS tools will have reduced value without targeted efforts to increase participation in research from populations across the world. Diversified genomic research, and comprehensive functional characterization of all variants, can ensure that pharmacogenomics guidelines, and CDS that are developed based on their recommendations, are inclusive.

## Unequal access to implement pharmacogenomic CDS resources

Implementing new and advanced technology in clinical settings is costly and requires specialized personnel, facility resources, and infrastructure. Large healthcare facilities are more likely to have the funding and expertise to implement new technologies and to educate their staff on how to correctly utilize them. In this way, pharmacogenomic CDS has the potential to increase healthcare disparities. The majority of pharmacogenomics research and implementation has been conducted at academic medical centers and large healthcare facilities, often excluding patients living in rural areas (Anderson et al., 2020; Dorfman et al., 2015; Duarte et al., 2021; Fohner et al., 2019; Hicks et al., 2021; Ramos et al., 2012). A CDS tool that caters to a large health system will likely not translate to smaller community-based health systems or facilities with fewer resources, where workflows or drug formularies may differ. For example, alternative medications recommended by the pharmacogenomic CDS tool may not be available in resource-limited settings. Furthermore, because specialty care may be more accessible in large healthcare settings, the cost/benefit of changing medications based on genotype versus using other interventions may vary. Therefore, even if the underlying genetic data are complete for all patients, pharmacogenomic CDS will likely be less beneficial for patients outside of large medical centers.

Community-based health systems, including those serving indigenous communities, are often the last to receive support in implementing innovative healthcare programs and, as such, they lag in their ability to implement pharmacogenomic CDS. Rural and tribal health systems are more likely to have technical limitations and lack of infrastructure to implement pharmacogenomics and may not have the capacity to store genetic information or to integrate with a larger health system to assist with pharmacogenomic CDS. Providers also may not have equal access to the most efficient or accurate genetic testing platforms. This sobering outlook can change, however. Research at University of Montana (UM) Skaggs Institute for Health Innovation (Box 4.1) provides a unique framework for implementing pharmacogenomics that serves underserved rural and tribal populations by leveraging telehealth and novel delivery modalities. The Northwest-Alaska Pharmacogenomics Research Network (Box 4.2) conducts community-engaged pharmacogenomics research with rural and tribal communities in the United States in order to broaden pharmacogenomics research in diverse populations and enhance the national environment for genomic research with participants from these communities (Boyer et al., 2011; Woodahl et al., 2014).

## Box 4.1 Skaggs Institute for Health Innovation (SIHI)

The Skaggs Institute for Health Innovation (SIHI) at the University of Montana Skaggs School of Pharmacy is a newly launched effort to ensure equitable access to pharmacogenomics for underserved communities (Skaggs Institute for Health Innovation, 2021). Advancements in precision medicine and pharmacogenomic testing have largely benefited patients living in proximity to large healthcare facilities or academic medical centers in major metropolitan areas, leaving many underserved populations behind. A priority of SIHI is to address healthcare disparities for rural and tribal patients who have historically been the last to gain access to emerging healthcare technologies, have less accessible medical care, and suffer from a higher burden of chronic disease. With over 44% of Montanans living in rural areas and a population density of 6.8 people per square mile, rurality poses unique challenges in providing healthcare. While Montana citizens are largely of European descent, American Indian people comprise the largest minority group (~7%) with 12 federally recognized tribes and 7 reservations in the state (Montana Indians: Their History and Location, 2021). University of Montana's SIHI is a centralized, statewide resource providing support to healthcare professionals, healthcare institutions, and patients toward implementation of pharmacogenomics.

We have partnered with healthcare systems across Montana to understand the unique challenges in implementing pharmacogenomics in resource-limited settings. This work has built on more than a decade of community-engaged research focusing on increasing participation of rural and tribal communities in precision medicine research. Key to these partnerships are building trust and working to ensure that the research addresses healthcare priorities of the community. We have worked with the Confederated Salish and Kootenai Tribes (CSKT) on the Flathead Reservation in a partnership that prioritizes community engagement to ensure tribal leaders, healthcare providers, and community members have input into precision medicine and pharmacogenetic research projects (Dorfman et al., 2015; Fohner et al., 2019; James et al., 2014; Morales et al., 2016; Woodahl et al., 2014). We have also partnered with Shodair Children's Hospital—the premier provider of psychiatric care to children across the state—and Partnership Health Center—a primary care center focused on providing equitable care for patients of all income levels. We have assessed provider and healthcare system needs for successful utilization of pharmacogenetic results as well as needs for integration into CDS tools. Across partner sites, there is enthusiasm for utilizing pharmacogenetic testing and increasing system-wide adoption. To facilitate access regardless of geographic location, pharmacogenomic consults will be delivered via telehealth and remote access. SIHI programs will help to ensure equitable access to pharmacogenomic testing for underserved communities in Montana and serve as a model for other resource-limited areas in the United States.

## Box 4.2 Northwest-Alaska Pharmacogenomics Research Network (NWA-PGRN)

The Pharmacogenomic Global Research Network (PGRN) is a collective of researchers with a focus on expanding information surrounding "precision medicine for the discovery and translation of genomic variation influencing therapeutic and adverse drug effects." A center funded by this network—the Northwest-Alaska Pharmacogenomics Research Network (NWA-PGRN)—focuses on expanding genomic research and access to pharmacogenetic testing for American Indian and Alaska Native (AIAN) people (Boyer et al., 2011; Woodahl et al., 2014).

AIAN people have historically been underrepresented in research, and this is especially true in genomics research. Researchers from Alaska, Montana, Oregon, and Washington have partnered with AIAN communities to ensure that the benefits emerging from the expansion of precision medicine and pharmacogenomics research will reach these communities. Central to the NWA-PGRN community-academic partnership is a focus on community-based participatory research to engage tribal communities in all aspects of the research. This includes the formation of community advisory boards, presentations to tribal leadership, gaining tribal approval for research projects, and working with the tribal community to develop research questions and procedures (Boyer et al., 2011; Morales et al., 2016; Woodahl et al., 2014). Community engagement has been essential to ensure that NWA-PGRN research is aligned with the health priorities of the community, is conducted in a culturally respectful manner, and that research results are disseminated.

NWA-PGRN research projects include pharmacogenetic association studies, gene–environment interactions, and qualitative research to better understand the unique obstacles and concerns AIAN people face when taking part in genomics research. Several projects have used a candidate gene approach to resequence pharmacogenes in order to characterize the full spectrum of variation in AIAN populations, including identifying variants not previously reported. This approach was taken to ensure that future pharmacogenetic testing platforms would include these unique variants providing improved clinically useful information for AIAN patients. One example is pharmacogenetic variation affecting warfarin dose and response in the Yup'ik people of western Alaska, where a high prevalence of a previously unreported variant in the *CYP2C9* gene was identified leading to a Met1Leu (M1L) amino acid change (Fohner et al., 2015). This variant results in non functional CYP2C9, leading to decreased metabolism of warfarin, and lower than average warfarin dose requirement for AIAN people in Alaska (Henderson et al., 2021; McDonald et al., 2020). Importantly, if testing platforms do not include the M1L variant, approximately 20% of Yup'ik people would be misclassified as being *CYP2C9*1/*1* based on incomplete genetic data. Another example identified a unique linkage disequilibrium

*(Continued)*

**Box 4.2 Northwest-Alaska Pharmacogenomics Research Network (NWA-PGRN) (*cont'd*)**

pattern in the *CYP3A4* and *CYP3A5* genes in the Confederated Salish and Kootenai Tribes (CSKT) in western Montana (Fohner et al., 2013). The *CYP3A4*1G* allele, which results in increased CYP3A4 activity, was observed at high frequency in CSKT people and patients with this allele may require higher doses of commonly prescribed medications metabolized by CYP3A4 (Fohner et al., 2021).

NWA-PGRN is founded on the principle that diversity and engagement in precision medicine is essential to ensuring health equity for underserved populations (Burke, 2021; Fohner et al., 2019). NWA-PGRN research highlights the importance of conducting pharmacogenetic research in underrepresented populations and not making assumptions about variation from previously published studies. By forming community—academic partnerships dedicated to increasing participation of AIAN people in pharmacogenomic research, NWA-PGRN hopes that the benefits offered by pharmacogenomics will be realized for these communities.

## Insurance coverage and financial resource challenges

Insurance coverage of pharmacogenomic testing varies widely. Historically, self-paying, affluent patients have had increased access to testing services, while insurance companies have been reluctant to reimburse for testing, citing concerns of uncertain clinical utility and cost (Keeling et al., 2019; Verbelen et al., 2017). Although the cost for pharmacogenomic testing is decreasing, it remains a financial burden for many patients and healthcare facilities. Cost-effectiveness research showing improved healthcare outcomes and cost savings resulting from pharmacogenomic testing may be an incentive for insurance companies to increase coverage for the cost of testing (Brixner et al., 2016; Empey et al., 2021; Groessl et al., 2018; Karamperis et al., 2021; Zhu et al., 2020). Additional data showing clinical utility are needed to show that investing in pharmacogenomics and pharmacogenomic CDS will improve health outcomes and decrease cost. Early adopters who widely share the benefits of pharmacogenomics in their health systems could encourage adoption by other health systems and also motivate insurance groups to expand access and reimbursement to more patients.

## Maximizing benefit and minimizing harm with pharmacogenomic CDS

With the aforementioned challenges, it is important to develop and implement pharmacogenomic CDS that maximizes benefits and minimizes harms for all patients. As research focuses on inclusion of diverse populations—across ethnicities, geography, age, disability, disease states, and socioeconomic status—the resulting pharmacogenomic CDS tools can help ensure health justice. To build these tools, providers and health systems should ideally use next-generation sequencing to capture all relevant pharmacogenomics variation in a given patient. Relying on targeted array-based genotyping platforms may not be useful for all patients because they include only a subset of variants that are representative of the global population.

To illustrate the value of this approach, consider the enzyme cytochrome P450 2D6 (CYP2D6), which controls the clearance of numerous drugs, including some with a narrow therapeutic window. The *CYP2D6* gene locus is highly variable with over 140 identified haplotypes that encode an enzyme with varying functionality and many with unknown or uncertain function (Gaedigk et al., 2019; Gaedigk et al., 2018). Targeted pharmacogenomics genotyping platforms include only a handful of these variants, potentially leaving out alleles that may have significant functional consequences, especially for non-European populations. Next-generation sequencing provides comprehensive identification of all relevant single nucleotide and structural variation in *CYP2D6* allowing CDS to offer a more accurate tool to benefit patients who otherwise may be classified as *CYP2D6*1/*1*, or normal metabolizers, using genotyping arrays (Dalton et al., 2020; Lee et al., 2019).

It is important to consider, however, that simply obtaining the sequencing data will be insufficient. Much more research is needed to characterize the function of novel and/or unknown function variants before they can be included in pharmacogenomic CDS with confident recommendations to change drug dose or choice. As is discussed in later chapters of this book, machine learning and deep mutational scanning offer novel methodologies to tackle the growing number of variants of unknown significance (Amorosi et al., 2021; Chiasson et al., 2020; McInnes et al., 2020; Suiter et al., 2020; van der Lee et al., 2021; Zhang et al., 2020, 2021).

Pharmacogenomic CDS tools must be transparent about which pharmacogenomic variants are tested, what other data each algorithm contains, and what resources were used to make clinical recommendations. This transparency will allow providers to determine whether the pharmacogenomic CDS-driven recommendation is applicable to a specific patient. It will also allow healthcare providers to evaluate how reliable and translatable results can be across patient cases. Pharmacogenomic CDS development should consider the challenges of implementation in clinics with smaller infrastructure. The tool should be tested and adapted to meet the needs of patients living in rural areas and those served by community-based health systems. Large health systems and CDS industries can facilitate effective implementation in smaller health systems by providing references to specialists and education for rural providers who may not have advanced resources. Telehealth consultations with pharmacogenomics specialists could give healthcare providers in diverse settings the ability to utilize pharmacogenomic CDS, helping to mitigate impediments related to education and location.

Once pharmacogenomic CDS is developed and implemented, drug treatment outcomes should be continuously monitored and integrated into the tool to improve its performance and recommendations, such as in a learning health system model. Pharmacogenomic CDS tools will need to be implemented prior to their perfection. As additional data on variants and drug response—especially in diverse populations—are collected, the pharmacogenomic CDS tool can be refined to include newly identified sources of variability and used to improve health equity. Assessments should be made about whether the tool decreases the time, effort, and resources required for providers in diverse settings, and how the tool could be optimized to improve its utility across these varied settings. Providers should have access to educational resources and specialists if there are conflicts or questions about the tool's process or utility, or how to interpret and use test results.

It is equally important to provide education and resources for the patients undergoing pharmacogenomic testing. Patients may not understand the test, or the benefits it provides, which means they may not seek follow-up or consultation about the results. Some patients are not aware that testing platforms vary and are not applicable for all medications. Another important consideration is the increase in direct-to-consumer pharmacogenomic testing platforms that may give patients unrealistic expectations about the benefits of genetic testing services (Gammal et al., 2021; Oh, 2019; Patel & McLeod, 2020).

Finally, it is necessary to clearly define the limitations of pharmacogenomic testing and to communicate that pharmacogenomic CDS is one of several tools to assist in medication therapy management. Providers should encourage patients to keep a record of their pharmacogenomic test results, and follow up as needed, as genetic information can be valuable in future medical decisions. It is important to recognize that some patients distrust or are wary of genetic testing, which may hinder uptake or willingness to consent to pharmacogenomic testing to inform treatment. This hesitancy is especially true of historically marginalized populations, who have endured a long history of injustices from research and medical programs, and who may be least likely to benefit from pharmacogenomics in its current form (Claw et al., 2018; Perera, 2019). The best way to overcome these sources of bias is by providing education on pharmacogenomic testing and healthcare privacy, building trust in the research community, and respecting patient autonomy should patients decline testing.

## Conclusion

Pharmacogenomic CDS has the potential to significantly increase access to cutting-edge healthcare innovations for underserved patients and to improve health outcomes as well as be a powerful tool to reduce healthcare disparities and improve equity across patient populations. Without proactive research and implementation strategies, however, pharmacogenomics can exacerbate health disparities rather than reduce them. Careful attention must be paid to existing inequities in representation of all people in research and implementation.

Pharmacogenomic CDS offers an exciting and promising route to improve the health of all patients regardless of race, ethnicity, geography, or class. By leveraging the unique genomic profile of individual patients, drugs can be optimized more quickly, precisely, and cheaply, without relying on crude racial categories and population averages. To realize the benefits of pharmacogenomic CDS and to contribute to health equity, we urge teams implementing pharmacogenomic CDS to embrace transparency and education, and to utilize a learning health system approach to continuously gather data and fill in existing gaps in pharmacogenomics knowledge. Through this approach, we can begin to implement pharmacogenomic CDS now, before tools are perfect, and can use the implementation process to overcome current limitations in pharmacogenomic data and CDS tools and to achieve equity in pharmacogenomics.

## References

Abul-Husn, N. S., & Kenny, E. E. (2019). Personalized medicine and the power of electronic health records. *Cell, 177*(1), 58–69. https://doi.org/10.1016/j.cell.2019.02.039

Amorosi, C. J., Chiasson, M. A., McDonald, M. G., Wong, L. H., Sitko, K. A., Boyle, G., Kowalski, J. P., Rettie, A. E., Fowler, D. M., & Dunham, M. J. (2021). Massively parallel characterization of CYP2C9 variant enzyme activity and abundance. *American Journal of Human Genetics, 108*(9), 1735–1751. https://doi.org/10.1016/j.ajhg.2021.07.001

Anderson, H. D., Crooks, K. R., Kao, D. P., & Aquilante, C. L. (2020). The landscape of pharmacogenetic testing in a US managed care population. *Genetics in Medicine, 22*(7), 1247–1253. https://doi.org/10.1038/s41436-020-0788-3

Bellon, T. (2021). *Bristol-Myers, Sanofi ordered to pay Hawaii $834 million over Plavix warning label.* https://www.reuters.com/article/us-bristol-myers-sanofi-plavix/bristol-myers-sanofi-ordered-to-pay-hawaii-834-million-over-plavix-warning-label-idUSKBN2AF1YI.

Botton, M. R., Whirl-Carrillo, M., Del Tredici, A. L., Sangkuhl, K., Cavallari, L. H., Agúndez, J. A. G., Duconge, J., Lee, M. T. M., Woodahl, E. L., Claudio-Campos, K., Daly, A. K., Klein, T. E., Pratt, V. M., Scott, S. A., & Gaedigk, A. (2021). PharmVar GeneFocus: CYP2C19. *Clinical Pharmacology and Therapeutics, 109*(2), 352–366. https://doi.org/10.1002/cpt.1973

Boyer, B. B., Dillard, D., Woodahl, E. L., Whitener, R., Thummel, K. E., & Burke, W. (2011). Ethical issues in developing pharmacogenetic research partnerships with American indigenous communities. *Clinical Pharmacology and Therapeutics, 89*(3), 343–345. https://doi.org/10.1038/clpt.2010.303

Brixner, D., Biltaji, E., Bress, A., Unni, S., Ye, X., Mamiya, T., Ashcraft, K., & Biskupiak, J. (2016). The effect of pharmacogenetic profiling with a clinical decision support tool on healthcare resource utilization and estimated costs in the elderly exposed to polypharmacy. *Journal of Medical Economics, 19*(3), 213–228. https://doi.org/10.3111/13696998.2015.1110160

Brown, J. T., Bishop, J. R., Sangkuhl, K., Nurmi, E. L., Mueller, D. J., Dinh, J., Gaedigk, A., Klein, T. E., Caudle, K. E., McCracken, J. T., de Leon, J., & Leeder, J. (2019). Clinical pharmacogenetics implementation consortium guideline for cytochrome P450 (CYP) 2D6 genotype and atomoxetine therapy. *Clinical Pharmacology and Therapeutics, 106*(1), 94–102. https://doi.org/10.1002/cpt.1409

Burke, Wylie (2021). Utility and diversity: Challenges for genomic medicine. *Annual Review of Genomics and Human Genetics, 22*, 1–24. https://doi.org/10.1146/annurev-genom-120220-082640

Burke, W., & Korngiebel, D. M. (2015). Closing the gap between knowledge and clinical application: Challenges for genomic translation. *PLoS Genetics, 11*(2), 1–9. https://doi.org/10.1371/journal.pgen.1004978

Caudle, K. E., Rettie, A. E., Whirl-Carrillo, M., Smith, L. H., Mintzer, S., Lee, M. T., Klein, T. E., & Callaghan, J. T. (2014). Clinical pharmacogenetics implementation consortium guidelines for CYP2C9 and HLA-B genotypes and phenytoin dosing. *Clinical Pharmacology and Therapeutics, 96*(5), 542–548. https://doi.org/10.1038/clpt.2014.159

Chiasson, M. A., Rollins, N. J., Stephany, J. J., Sitko, K. A., Matreyek, K. A., Verby, M., Sun, S., Roth, F. P., De Sloover, D., Marks, D. S., Rettie, A. E., & Fowler, D. M. (2020). Multiplexed measurement of variant abundance and activity reveals VKOR topology, active site and human variant impact. *eLife, 9*, 1–25. https://doi.org/10.7554/ELIFE.58026

Claw, K. G., Anderson, M. Z., Begay, R. L., Tsosie, K. S., Fox, K., Garrison, N. A., Bader, A. C., Bardill, J., Bolnick, D. A., Brooks, J., Cordova, A., Malhi, R. S., Nakatsuka, N., Neller, A., Raff, J. A., Singson, J., TallBear, K., Vargas, T., &

Yracheta, J. M. (2018). A framework for enhancing ethical genomic research with Indigenous communities. *Nature Communications, 9*(1). https://doi.org/10.1038/s41467-018-05188-3

Dalton, R., Lee, S.b., Claw, K. G., Prasad, B., Phillips, B. R., Shen, D. D., Wong, L. H., Fade, M., McDonald, M. G., Dunham, M. J., Fowler, D. M., Rettie, A. E., Schuetz, E., Thornton, T. A., Nickerson, D. A., Gaedigk, A., Thummel, K. E., & Woodahl, E. L. (2020). Interrogation of CYP2D6 structural variant alleles improves the correlation between CYP2D6 genotype and CYP2D6-mediated metabolic activity. *Clinical and Translational Science, 13*(1), 147–156. https://doi.org/10.1111/cts.12695

Dang, M. T. N., Hambleton, J., & Kayser, S. R. (2005). The influence of ethnicity on warfarin dosage requirement. *Annals of Pharmacotherapy, 39*(6), 1008–1012. https://doi.org/10.1345/aph.1E566

Dorfman, E. H., Brown Trinidad, S., Morales, C. T., Howlett, K., Burke, W., & Woodahl, E. L. (2015). Pharmacogenomics in diverse practice settings: Implementation beyond major metropolitan areas. *Pharmacogenomics, 16*(3), 227–237. https://doi.org/10.2217/pgs.14.174

Duarte, J. D., Dalton, R., Elchynski, A. L., Smith, D. M., Cicali, E. J., Lee, J. C., Duong, B. Q., Petry, N. J., Aquilante, C. L., Beitelshees, A. L., Empey, P. E., Johnson, J. A., Obeng, A. O., Pasternak, A. L., Pratt, V. M., Ramsey, L. B., Tuteja, S., Van Driest, S. L., Wiisanen, K., … Cavallari, L. H. (2021). Multisite investigation of strategies for the clinical implementation of pre-emptive pharmacogenetic testing. *Genetics in Medicine, 23*(12), 2335–2341. https://doi.org/10.1038/s41436-021-01269-9

Empey, P. E., Pratt, V. M., Hoffman, J. M., Caudle, K. E., & Klein, T. E. (2021). Expanding evidence leads to new pharmacogenomics payer coverage. *Genetics in Medicine, 23*(5), 830–832. https://doi.org/10.1038/s41436-021-01117-w

Fohner, A. E., Dalton, R., Skagen, K., Jackson, K., Claw, K. G., Hopkins, S. E., Robinson, R., Khan, B. A., Prasad, B., Schuetz, E. G., Nickerson, D. A., Thornton, T. A., Dillard, D. A., Boyer, B. B., Thummel, K. E., & Woodahl, E. L. (2021). Characterization of CYP3A pharmacogenetic variation in American Indian and Alaska Native communities, targeting CYP3A4*1G allele function. *Clinical and Translational Science, 14*(4), 1292–1302. https://doi.org/10.1111/cts.12970

Fohner, A., Muzquiz, L. I., Austin, M. A., Gaedigk, A., Gordon, A., Thornton, T., Rieder, M. J., Pershouse, M. A., Putnam, E. A., Howlett, K., Beatty, P., Thummel, K. E., & Woodahl, E. L. (2013). Pharmacogenetics in American Indian populations: analysis of CYP2D6, CYP3A4, CYP3A5, and CYP2C9 in the Confederated Salish and Kootenai Tribes. *Pharmacogenet Genomics, 23*(8), 403–414. https://doi.org/10.1097/FPC.0b013e3283629ce9. PMID: 23778323; PMCID: PMC3937472.

Fohner, A. E., Robinson, R., Yracheta, J., Dillard, D. A., Schilling, B., Khan, B., Hopkins, S., Boyer, B., Black, J., Wiener, H., Tiwari, H. K., Gordon, A., Nickerson, D., Tsai, J. M., Farin, F. M., Thornton, T. A., Rettie, A. E., & Thummel, K. E. (2015). Variation in genes controlling warfarin disposition and response in American Indian and Alaska native people: CYP2C9, VKORC1, CYP4F2, CYP4F11, GGCX. *Pharmacogenetics and Genomics, 25*(7), 343–353. https://doi.org/10.1097/FPC.0000000000000143

Fohner, A. E., Volk, K. G., & Woodahl, E. L. (2019). Democratizing precision medicine through community engagement. *Clinical Pharmacology and Therapeutics, 106*(3), 488–490. https://doi.org/10.1002/cpt.1508

Gaedigk, Andrea, Ingelman-Sundberg, M., Miller, N. A., Leeder, J. S., Whirl-Carrillo, M., & Klein, T. E. (2018). The pharmacogene variation (PharmVar) consortium:

Incorporation of the human cytochrome P450 ( CYP ) allele nomenclature database. *Clinical Pharmacology and Therapeutics, 103*(3), 399–401. https://doi.org/10.1002/cpt.910

Gaedigk, A., Sangkuhl, K., Whirl-Carrillo, M., Twist, G. P., Klein, T. E., & Miller, N. A. (2019). The evolution of PharmVar. *Clinical Pharmacology and Therapeutics, 105*(1), 29–32. https://doi.org/10.1002/cpt.1275

Gammal, R. S., Smith, D. M., Wiisanen, K. W., Cusimano, J. M., Pettit, R. S., Stephens, J. W., Walko, C. M., Yang, R. H., & Dunnenberger, H. M. (2021). The pharmacist's responsibility to ensure appropriate use of direct-to-consumer genetic testing. *JACCP Journal of the American College of Clinical Pharmacy, 4*(5), 652–658. https://doi.org/10.1002/jac5.1437

Gladding, P., MacKay, J., Zeng, I., Stewart, R., Prabkahar, R., Webster, M., & White, H. (2010). A simulation of warfarin maintenance dose requirement using a pharmacogenetic algorithm in an ethnically diverse cohort. *Personalized Medicine,* 7(3), 319–325. https://doi.org/10.2217/pme.10.24

Groessl, E. J., Tally, S. R., Hillery, N., Maciel, A., & Garces, J. A. (2018). Cost-effectiveness of a pharmacogenetic test to guide treatment for major depressive disorder. *Journal of Managed Care and Specialty Pharmacy, 24*(8), 726–734. https://doi.org/10.18553/jmcp.2018.24.8.726

Henderson, L. M., Hopkins, S. E., Boyer, B. B., Thornton, T. A., Rettie, A. E., & Thummel, K. E. (2021). A novel and common variant in the Yup'Ik Alaska native population. *Drug Metabolism and Disposition: The Biological Fate of Chemicals, 49*(5), 345–352. https://doi.org/10.1124/dmd.120.000301

Hershfield, M. S., Callaghan, J. T., Tassaneeyakul, W., Mushiroda, T., Thorn, C. F., Klein, T. E., & Lee, M. T. M. (2013). Clinical pharmacogenetics implementation consortium guidelines for human leukocyte antigen-b genotype and allopurinol dosing. *Clinical Pharmacology and Therapeutics, 93*(2), 153–158. https://doi.org/10.1038/clpt.2012.209

Hicks, J., Bishop, J., Sangkuhl, K., Müller, D., Ji, Y., Leckband, S., Leeder, J., Graham, R., Chiulli, D., LLerena, A., Skaar, T., Scott, S., Stingl, J., Klein, T., Caudle, K., & Gaedigk, A. (2015). Clinical pharmacogenetics implementation consortium (CPIC) guideline for CYP2D6 and CYP2C19 genotypes and dosing of selective serotonin reuptake inhibitors. *Clinical Pharmacology and Therapeutics, 98*(2), 127–134. https://doi.org/10.1002/cpt.147

Hicks, J. K., El Rouby, N., Ong, H. H., Schildcrout, J. S., Ramsey, L. B., Shi, Y., Anne Tang, L., Aquilante, C. L., Beitelshees, A. L., Blake, K. V., Cimino, J. J., Davis, B. H., Empey, P. E., Kao, D. P., Lemkin, D. L., Limdi, N. A., Lipori, G. P., Rosenman, M. B., Skaar, T. C., … Peterson, J. F. (2021). Opportunity for genotype-guided prescribing among adult patients in 11 US health systems. *Clinical Pharmacology and Therapeutics, 110*(1), 179–188. https://doi.org/10.1002/cpt.2161

Hicks, J., Sangkuhl, K., Swen, J., Ellingrod, V., Müller, D., Shimoda, K., Bishop, J., Kharasch, E., Skaar, T., Gaedigk, A., Dunnenberger, H., Klein, T., Caudle, K., & Stingl, J. (2017). Clinical pharmacogenetics implementation consortium guideline (CPIC) for CYP2D6 and CYP2C19 genotypes and dosing of tricyclic antidepressants: 2016 update. *Clinical Pharmacology and Therapeutics, 102*(1), 37–44. https://doi.org/10.1002/cpt.597

Huddart, R., Fohner, A. E., Whirl-Carrillo, M., Wojcik, G. L., Gignoux, C. R., Popejoy, A. B., Bustamante, C. D., Altman, R. B., & Klein, T. E. (2019). Standardized biogeographic grouping system for annotating populations in pharmacogenetic research. *Clinical Pharmacology and Therapeutics, 105*(5), 1256–1262. https://doi.org/10.1002/cpt.1322

Hunt, L. M., & Kreiner, M. J. (2013). Pharmacogenetics in primary care: The promise of personalized medicine and the reality of racial profiling. *Culture, Medicine and Psychiatry, 37*(1), 226–235. https://doi.org/10.1007/s11013-012-9303-x

James, R., Tsosie, R., Sahota, P., Parker, M., Dillard, D., Sylvester, I., Lewis, J., Klejka, J., Muzquiz, L., Olsen, P., Whitener, R., & Burke, W. (2014). Exploring pathways to trust: A tribal perspective on data sharing. *Genetics in Medicine, 16*(11), 820–826. https://doi.org/10.1038/gim.2014.47

Johnson, J. A., Caudle, K. E., Gong, L., Whirl-Carrillo, M., Stein, C. M., Scott, S. A., Lee, M. T., Gage, B. F., Kimmel, S. E., Perera, M. A., Anderson, J. L., Pirmohamed, M., Klein, T. E., Limdi, N. A., Cavallari, L. H., & Wadelius, M. (2017). Clinical pharmacogenetics implementation consortium (CPIC) guideline for pharmacogenetics-guided warfarin dosing: 2017 update. *Clinical Pharmacology and Therapeutics, 102*(3), 397–404. https://doi.org/10.1002/cpt.668

Johnson, J. A., Gong, L., Whirl-Carrillo, M., Gage, B. F., Scott, S. A., Stein, C. M., Anderson, J. L., Kimmel, S. E., Lee, M. T. M., Pirmohamed, M., Wadelius, M., Klein, T. E., & Altman, R. B. (2011). Clinical pharmacogenetics implementation consortium guidelines for CYP2C9 and VKORC1 genotypes and warfarin dosing. *Clinical Pharmacology and Therapeutics, 90*(4), 625–629. https://doi.org/10.1038/clpt.2011.185

Karamperis, K., Koromina, M., Papantoniou, P., Skokou, M., Kanellakis, F., Mitropoulos, K., Vozikis, A., Müller, D. J., Patrinos, G. P., & Mitropoulou, C. (2021). Economic evaluation in psychiatric pharmacogenomics: A systematic review. *The Pharmacogenomics Journal, 21*(4), 533–541. https://doi.org/10.1038/s41397-021-00249-1

Karnes, J. H., Rettie, A. E., Somogyi, A. A., Huddart, R., Fohner, A. E., Formea, C. M., Ta Michael Lee, M., Llerena, A., Whirl-Carrillo, M., Klein, T. E., Phillips, E. J., Mintzer, S., Gaedigk, A., Caudle, K. E., & Callaghan, J. T. (2021). Clinical pharmacogenetics implementation consortium (CPIC) guideline for CYP2C9 and HLA-B genotypes and phenytoin dosing: 2020 update. *Clinical Pharmacology and Therapeutics, 109*(2), 302–309. https://doi.org/10.1002/cpt.2008

Kaye, J. B., Schultz, L. E., Steiner, H. E., Kittles, R. A., Cavallari, L. H., & Karnes, J. H. (2017). Warfarin pharmacogenomics in diverse populations. *Pharmacotherapy, 37*(9), 1150–1163. https://doi.org/10.1002/phar.1982

Keeling, N. J., Rosenthal, M. M., West-Strum, D., Patel, A. S., Haidar, C. E., & Hoffman, J. M. (2019). Preemptive pharmacogenetic testing: Exploring the knowledge and perspectives of US payers. *Genetics in Medicine, 21*(5), 1224–1232. https://doi.org/10.1038/gim.2017.181

Lam, M. P. S., & Cheung, B. M. Y. (2012). The pharmacogenetics of the response to warfarin in Chinese. *British Journal of Clinical Pharmacology, 73*(3), 340–347. https://doi.org/10.1111/j.1365-2125.2011.04097.x

Leckband, S. G., Kelsoe, J. R., Dunnenberger, H. M., George, A. L., Tran, E., Berger, R., Müller, D. J., Whirl-Carrillo, M., Caudle, K. E., & Pirmohamed, M. (2013). Clinical pharmacogenetics implementation consortium guidelines for HLA-B genotype and carbamazepine dosing. *Clinical Pharmacology and Therapeutics, 94*(3), 324–328. https://doi.org/10.1038/clpt.2013.103

van der Lee, M., Allard, W. G., Vossen, R. H. A. M., Baak-Pablo, R. F., Menafra, R., Deiman, B. A. L. M., Deenen, M. J., Neven, P., Johansson, I., Gastaldello, S., Ingelman-Sundberg, M., Guchelaar, H.-J., Swen, J. J., & Anvar, S. Y. (2021). Toward predicting CYP2D6-mediated variable drug response from CYP2D6 gene sequencing data. *Science Translational Medicine, 13*(603). https://doi.org/10.1126/scitranslmed.abf3637

Lee, S.b., Wheeler, M. M., Patterson, K., McGee, S., Dalton, R., Woodahl, E. L., Gaedigk, A., Thummel, K. E., & Nickerson, D. A. (2019). Stargazer: A software tool for calling star alleles from next-generation sequencing data using CYP2D6 as a model. *Genetics in Medicine, 21*(2), 361–372. https://doi.org/10.1038/s41436-018-0054-0

Luczak, T., Stenehjem, D., & Brown, J. (2021). Applying an equity lens to pharmacogenetic research and translation to under-represented populations. *Clinical and Translational Science, 14*(6), 2117–2123. https://doi.org/10.1111/cts.13110

Magavern, E. F., Gurdasani, D., Ng, F. L., & Lee, S. S. (2022). Health equality, race and pharmacogenomics. *British Journal of Clinical Pharmacology, 88*(1), 27–33. https://doi.org/10.1111/bcp.14983. Epub 2021 Aug 4. PMID: 34251046; PMCID: PMC8752640.

Martin, A., Downing, J., Maden, M., Fleeman, N., Alfirevic, A., Haycox, A., & Pirmohamed, M. (2017). An assessment of the impact of pharmacogenomics on health disparities: A systematic literature review. *Pharmacogenomics, 18*(16), 1541–1550. https://doi.org/10.2217/pgs-2017-0076

Martin, M. A., Hoffman, J. M., Freimuth, R. R., Klein, T. E., Dong, B. J., Pirmohamed, M., Hicks, J. K., Wilkinson, M. R., Haas, D. W., & Kroetz, D. L. (2014). Clinical pharmacogenetics implementation consortium guidelines for HLA-B genotype and abacavir dosing: 2014 update. *Clinical Pharmacology and Therapeutics, 95*(5), 499–500. https://doi.org/10.1038/clpt.2014.38

McDonald, M. G., Henderson, L. M., Ray, S., Yeung, C. K., Johnson, A. L., Kowalski, J. P., Hanenberg, H., Wiek, C., Thummel, K. E., & Rettie, A. E. (2020). Heterologous expression and functional characterization of novel CYP2C9 variants identified in the Alaska native people. *The Journal of Pharmacology and Experimental Therapeutics, 374*(2), 233–240. https://doi.org/10.1124/jpet.120.265850

McInnes, G., Dalton, R., Sangkuhl, K., Whirl-Carrillo, M., Lee, S., Tsao, P. S., Gaedigk, A., Altman, R. B., Woodahl, E. L., & Schlessinger, A. (2020). Transfer learning enables prediction of CYP2D6 haplotype function. *PLOS Computational Biology, 16*(11), e1008399. https://doi.org/10.1371/journal.pcbi.1008399

*Montana Indians: Their history and location.*(2021). https://opi.mt.gov/Portals/182/Page%20Files/Indian%20Education/Indian%20Education%20101/Montana%20Indians%20Their%20History%20and%20Location.pdf.

Morales, C. T., Muzquiz, L. I., Howlett, K., Azure, B., Bodnar, B., Finley, V., Incashola, T., Mathias, C., Laukes, C., Beatty, P., Burke, W., Pershouse, M. A., Putnam, E. A., Trinidad, S. B., James, R., & Woodahl, E. L. (2016). Partnership with the confederated salish and Kootenai tribes: Establishing an advisory committee for pharmacogenetic research. *Progress in Community Health Partnerships : Research, Education, and Action, 10*(2), 173–183. https://doi.org/10.1353/cpr.2016.0035

Oh, B. (2019). Direct-to-consumer genetic testing: Advantages and pitfalls. *Genomics & Informatics, 17*(3), e33. https://doi.org/10.5808/gi.2019.17.3.e33

Patel, J. N., & McLeod, H. L. (2020). Clarity also needed for direct-to-consumer pharmacogenetic tests. *Journal of Clinical Oncology, 38*(3), 283–284. https://doi.org/10.1200/JCO.19.01917

Perera, M. (2019). Leaving some behind: The growing gap in precision medicine for minority populations. *Expert Review of Precision Medicine and Drug Development, 4*(1), 3–5. https://doi.org/10.1080/23808993.2019.1529541

Perera, M. A., Cavallari, L. H., & Johnson, J. A. (2014). Warfarin pharmacogenetics: An illustration of the importance of studies in minority populations. *Clinical Pharmacology and Therapeutics, 95*(3), 242–244. https://doi.org/10.1038/clpt.2013.209

Phillips, E. J., Sukasem, C., Whirl-Carrillo, M., Müller, D. J., Dunnenberger, H. M., Chantratita, W., Goldspiel, B., Chen, Y. T., Carleton, B. C., George, A. L., Mushiroda, T., Klein, T., Gammal, R. S., & Pirmohamed, M. (2018). Clinical pharmacogenetics implementation consortium guideline for HLA genotype and use of carbamazepine and oxcarbazepine: 2017 update. *Clinical Pharmacology and Therapeutics, 103*(4), 574–581. https://doi.org/10.1002/cpt.1004

Popejoy, A. B., & Fullerton, S. M. (2016). Genomics is failing on diversity. *Nature, 538*(7624), 161–164. https://doi.org/10.1038/538161a

Ramos, E., Callier, S. L., & Rotimi, C. N. (2012). Why personalized medicine will fail if we stay the course. *Personalized Medicine, 9*(8), 839–847. https://doi.org/10.2217/pme.12.100

Roberts, D. (2011). *Fatal invention: How science, politics, and big business Re-create race in the twenty-first century*.

Saini, A. (2019). *Superior: The return of race science*.

Saito, Y., Stamp, L., Caudle, K., Hershfield, M., McDonagh, E., Callaghan, J., Tassaneeyakul, W., Mushiroda, T., Kamatani, N., Goldspiel, B., Phillips, E., Klein, T., & Lee, M. (2016). Clinical pharmacogenetics implementation consortium (CPIC) guidelines for human leukocyte antigen B (HLA-B) genotype and allopurinol dosing: 2015 update. *Clinical Pharmacology and Therapeutics, 99*(1), 36–37. https://doi.org/10.1002/cpt.161

Scott, S. A., Sangkuhl, K., Gardner, E. E., Stein, C. M., Hulot, J. S., Johnson, J. A., Roden, D. M., Klein, T. E., & Shuldiner, A. R. (2011). Clinical pharmacogenetics implementation consortium guidelines for cytochrome P450-2C19 (CYP2C19) genotype and clopidogrel therapy. *Clinical Pharmacology and Therapeutics, 90*(2), 328–332. https://doi.org/10.1038/clpt.2011.132

Scott, S. A., Sangkuhl, K., Stein, C. M., Hulot, J. S., Mega, J. L., Roden, D. M., Klein, T. E., Sabatine, M. S., Johnson, J. A., & Shuldiner, A. R. (2013). Clinical pharmacogenetics implementation consortium guidelines for CYP2C19 genotype and clopidogrel therapy: 2013 update. *Clinical Pharmacology and Therapeutics, 94*(3), 317–323. https://doi.org/10.1038/clpt.2013.105

Skaggs Institute for Health Innovation. (2021). https://umt.edu/pgx.

Suiter, C. C., Moriyama, T., Matreyek, K. A., Yang, W., Scaletti, E. R., Nishii, R., Yang, W., Hoshitsuki, K., Singh, M., Trehan, A., Parish, C., Smith, C., Li, L., Bhojwani, D., Yuen, L. Y. P., Li, C., Li, C., Yang, Y., Walker, G. J., … Yang, J. J. (2020). Massively parallel variant characterization identifies NUDT15 alleles associated with thiopurine toxicity. *Proceedings of the National Academy of Sciences, 117*(10), 5394–5401. https://doi.org/10.1073/pnas.1915680117

Verbelen, M., Weale, M. E., & Lewis, C. M. (2017). Cost-effectiveness of pharmacogenetic-guided treatment: Are we there yet? *The Pharmacogenomics Journal, 17*(5), 395–402. https://doi.org/10.1038/tpj.2017.21

Wake, D. T., Smith, D. M., Kazi, S., & Dunnenberger, H. M. (2021). Pharmacogenomic clinical decision support: A review, how-to guide, and future vision. *Clinical Pharmacology and Therapeutics*. https://doi.org/10.1002/cpt.2387. Epub ahead of print. PMID: 34365648.

Whirl-Carrillo, M., Huddart, R., Gong, L., Sangkuhl, K., Thorn, C. F., Whaley, R., & Klein, T. E. (2021). An evidence-based framework for evaluating pharmacogenomics knowledge for personalized medicine. *Clinical Pharmacology and Therapeutics, 110*(3), 563–572. https://doi.org/10.1002/cpt.2350

Woodahl, E. L., Lesko, L. J., Hopkins, S., Robinson, R. F., Thummel, K. E., & Burke, W. (2014). Pharmacogenetic research in partnership with American Indian and Alaska native communities. *Pharmacogenomics, 15*(9), 1235–1241. https://doi.org/10.2217/pgs.14.91

Yang, L., Ge, W., Yu, F., & Zhu, H. (2010). Impact of VKORC1 gene polymorphism on interindividual and interethnic warfarin dosage requirement- A systematic review and meta analysis. *Thrombosis Research, 125*(4), e159–e166. https://doi.org/10.1016/j.thromres.2009.10.017

Zhang, L., Sarangi, V., Ho, M. F., Moon, I., Kalari, K. R., Wang, L., & Weinshilboum, R. M. (2021). SLCO1B1: Application and limitations of deep mutational scanning for genomic missense variant function. *Drug Metabolism and Disposition, 49*(5), 395–404. https://doi.org/10.1124/dmd.120.000264

Zhang, L., Sarangi, V., Moon, I., Yu, J., Liu, D., Devarajan, S., Reid, J. M., Kalari, K. R., Wang, L., & Weinshilboum, R. (2020). CYP2C9 and CYP2C19: Deep mutational scanning and functional characterization of genomic missense variants. *Clinical and Translational Science, 13*(4), 727–742. https://doi.org/10.1111/cts.12758

Zhu, Y., Swanson, K. M., Rojas, R. L., Wang, Z., St Sauver, J. L., Visscher, S. L., Prokop, L. J., Bielinski, S. J., Wang, L., Weinshilboum, R., & Borah, B. J. (2020). Systematic review of the evidence on the cost-effectiveness of pharmacogenomics-guided treatment for cardiovascular diseases. *Genetics in Medicine, 22*(3), 475–486. https://doi.org/10.1038/s41436-019-0667-y

SECTION TWO

# Developing and building clinical decision support for pharmacogenomic precision medicine

CHAPTER FIVE

# Clinical decision support methods and infrastructure

**Nephi Walton[1], Bret Heale[1], Christine Formea[1,2]**
[1]Intermountain Healthcare, Intermountain Precision Genomics, Salt Lake City, UT, United States
[2]Intermountain Department of Pharmacy, Intermountain Precision Genomics, Salt Lake City, UT, United States

## Contents



## Introduction

When considering clinical decision support (CDS) methods and infrastructure for pharmacogenomics, an institution may opt to use the native electronic health record (EHR) functionality, build custom tools within the native EHR, or purchase a third-party application. There are advantages and drawbacks to each of these options. The decision should largely be based on the degree of customization needed by the organization and its willingness to dedicate resources to maintain both the technical and knowledge-

*Clinical Decision Support for Pharmacogenomic Precision Medicine*
ISBN: 978-0-12-824453-1
https://doi.org/10.1016/B978-0-12-824453-1.00001-4

based components of the implementation. Additionally, it is important that the chosen solution is concordant with the institutional target for pharmacogenomics. An advantage of third-party tools is that the maintenance of the knowledge base and CDS rules are independent from the institution. Also, vendors may offer an environment that will support simple customization of rules while enabling nontechnical pharmacogenomics experts to maintain the CDS. However, there is also the potential drawback that the tools might not offer CDS rules and methods suitable for the organization. This chapter reviews modes of information delivery in CDS and options for developing pharmacogenomic CDS, including within a native EHR system or through a third-party application.

## Modes of information delivery in CDS

### Push versus pull strategies

A CDS strategy can deliver information via a *push* or *pull* mode of delivery. In the push method, decision support is pushed to the provider in the clinical workflow regardless of whether they are seeking it. This tends to be active decision support, although it can also be passive if it is not interrupting the clinical workflow. Both active and passive CDS are described further below. The pull modality of CDS requires a provider to seek out knowledge or support and is more often employed when a provider is actively seeking information that is required to care for a patient. The primary disadvantage of a pull approach with pharmacogenomics in its current state is that providers generally do not seek out pharmacogenomic guidance prior to dosing medications. As the number of patients with pharmacogenomic testing increases, and guidance on using pharmacogenomics in medication prescribing becomes clearer, the information seeking by clinicians will become more commonplace, leading to increased effectiveness of pull methods for CDS delivery.

### Passive versus active CDS

*Passive* decision support consists of several different modalities. This type of CDS can still be structured as an alert but not necessarily one that has to be clicked on in order to move through the workflow. For example, when placing a medication order, an alert with a red exclamation point can appear on the ordering screen while a clinician is ordering a medication that needs to be dose adjusted due to a specific metabolizer status. This alert would be informational only, not requiring any physician interaction, and would disappear when the clinician completes the order. There may be passive CDS accompanying the alert that requires action by the physician to access recommendations. For example, clicking on a link to a knowledge source in a results display or dashboard. There are many different ways passive alerts can manifest, including but not limited to, integrated result displays, dashboards, or Infobuttons. Infobuttons, described in greater detail later in this chapter, can provide access to external knowledge sources for point-of-care healthcare providers in a just-in-time manner (Heale et al., 2016; Williams et al., 2019). SMART apps, also described later, can provide highly contextualized and actionable decision support using health information technology standards (Alterovitz et al., 2015; Mandel et al., 2016). If not designed appropriately, even passive CDS alerts may result in excessive low-priority clinical alerts and ultimately cause healthcare provider fatigue and make their utility less effective (Keller, 2012; Kizzier-Carnahan et al., 2019).

In contrast, *active* decision support involves CDS that interrupts the clinical workflow to prompt the provider to make a change. A distinguishing feature is that active decision support forces the provider to acknowledge the existence of the warning before they can move on in the patient care workflow. This decision support modality tends to be the most effective in changing prescriber behavior (Bell et al., 2014). However, it is also the most disruptive. If not deployed strategically, active decision support can cause alert fatigue resulting in providers dismissing alerts without reading or fully understanding them (Keller, 2012; Kizzier-Carnahan et al., 2019).

Active decision support is most useful when the impact of prescribing the wrong medication is high and the occurrence of the alert is rare. Passive decision support is better used when the impact is not as high, the occurrence is more common, and where there are multiple clinical variables involved in prescribing that may easily override the importance of the pharmacogenomics result. Passive decision support is more useful when prescribers are using the pharmacogenomic data to troubleshoot concerns to rule out genetic interactions, such as in preordering a test prior to prescribing a medication.

## Synchronous versus asynchronous CDS

*Synchronous* decision support is a strategy in which the alert is well aligned with the clinical workflow and is placed where the prescriber is actively prescribing a medication that might be affected by CDS. If synchronous alerting is desired, the CDS infrastructure must support real-time interaction between the clinical system, the recommendation, and the logic that triggers the alert. While many EHRs are equipped with integrated alert triggering components that could support this interaction, genetic data are often better supported outside of the EHR. In this case, outsourcing pharmacogenomics CDS to third-party or custom applications outside of the EHR provides another option. However, a key component of the alert is the message. If adopting solutions that reside outside of the EHR, additional architecture is needed to bring the message to the EHR, unless the choice is made to store recommendations within the EHR as static documents.

In *asynchronous* decision support, the alert is active but out of step with clinical workflow, attempting to get the information to the provider (or patient) even before they have a decision to make. One example of asynchronous CDS for pharmacogenomics is alerting the provider that a pharmacogenomics test report is available for the patient once the report is stored in the EHR. Another example is placing a potential pharmacogenomics interaction in the problem list, allergy list, or patient chart "adjacent information zone" (e.g., Cerner SmartZone). Such alerts may also be presented to healthcare providers in an aggregate manner, such as when new pharmacogenomics knowledge becomes available or batches of patients require data updates for quality purposes (Williams et al., 2019). Other examples may include patient-level alerts for new information on genetic variants and population health level evaluations for patients meeting criteria for genomic care (Williams et al., 2019).

There are many ways pharmacogenomic data can drive CDS, as illustrated in Fig. 5.1. An ideal pharmacogenomics implementation would be comprised of a mix of alerts and allow for strategic customization to optimize utility and minimize disruption (Hinderer et al., 2017). Ultimately the decision to use one or more of these modality is based on institutional capabilities and preferences and the purpose of the alert (e.g., informative vs. risk aversion).

## Vendor-specific electronic health record systems CDS capabilities

The specific technical capabilities of a given vendor-specific EHR system may limit CDS implementations. While some major EHR vendors have developed support for structured genetic data, enabling the ability to apply CDS logic, most data are currently stored in a PDF format that is

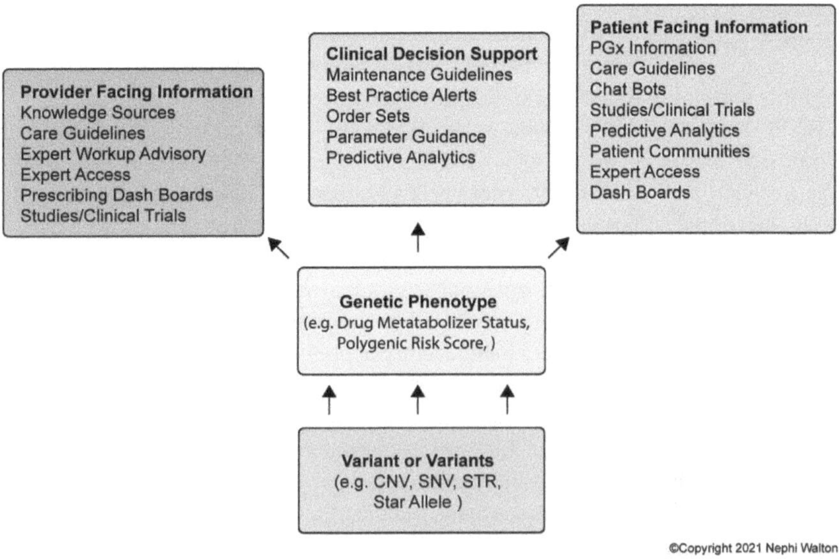


**Figure 5.1** Diagram demonstrating pathway from variant, to phenotype, to various use cases of pharmacogenomic data. An institution's decision to use one or more CDS modes of delivery is largely based on institutional capabilities and preferences and the purpose of the alert.

not readable by the EHR system. This limitation presents a significant challenge when developing CDS for pharmacogenomics or other genetic data.

Transfer of pharmacogenomic data from the laboratory into the EHR can pose challenges. A custom interface within the EHR is required from the laboratory in order to transmit genetic data into the EHR. Alternatively, manual entry of the genetic data is not a realistic expectation in the clinical setting and poses a significant risk of human error. Most successful deployments of pharmacogenomic CDS in native EHR systems are based on the use of a single pharmacogenomics laboratory for which a health system has built a custom data transfer interface. The availability of standardized interfaces for writing pharmacogenomic lab results to the patient chart will expand as EHR vendors adopt standards for representing pharmacogenomic lab results in the health record and standards for exchanging pharmacogenomic laboratory information become more mature.

While multiple EHR vendors have varying capabilities to manage pharmacogenomics, two of the largest vendors used for pharmacogenomics are Cerner and Epic. Boxes 5.1 and 5.2 present authors' experiences with each these vendors to illustrate practical issues with CDS capabilities. This

## Box 5.1 Native EHR capabilities: Cerner example

Cerner authored a paper illustrating some of the challenges with pharmacogenomics and their vision for a standards-based approach to pharmacogenomics within the EHR in 2019 (Collins et al., 2019). The paper presented a practical approach that is yet to be fully realized. One aspect to EHR CDS is how the data driving that CDS arrive. Cerner's EHR architecture is based on observed results. Currently in Cerner, while it is possible to build event codes to capture biomarkers, each biomarker needs to be "purpose built" and is specific to a result. Unfortunately, this approach does not scale to large numbers of pharmacogenomic interactions and requires continual maintenance to be updated. Cerner's architecture can accommodate manual entry of potential pharmacogenomics interactions into problem or allergy lists as well as the automatic population of problem lists when triggering CDS. For whole genome sequencing data where variant data are not ephemeral but the interpretation is, one could seek a Cerner build that pulls out specific biomarkers (traditional variant interpreted with context) and places them into purpose-built centralized event codes. However, this runs into a scaling issue. Capture of the sequence data elsewhere or in a new structure would lend itself to CDS that can scale or move the maintenance of rules outside the local team and Cerner.

To place important pharmacogenomic content in front of clinicians without disrupting workflow, some additional landing places other than alerts are offered by Cerner. The Cerner Smart-Zone provides a specialized niche area within EHR user interface to provide a less obtrusive landing place for CDS recommendations. Additionally, custom Millennium pages ("mpages"), views inside the EHR, can be created to provide a place for passive alerting. The Smart-Zone and custom mpages have advantages compared with adding pharmacogenomics interactions to the allergy list but would not be ideal for life-threatening interactions. Finally, Cerner has been made progress using the CDS Hooks standard (https://cds-hooks.hl7.org/) to deliver pharmacogenomics, but this functionality is limited to a synthetic "sandbox" environment.

discussion is not intended as an endorsement of either vendor but rather a discussion of how each vendor currently manages pharmacogenomics at the time of writing this chapter.

EHR vendors are in the early stages of deployment of architectures to manage genomic data. As with any new technologies, there are challenges in their initial deployment and the systems will mature and improve over time. Whether any given vendor's technology is mature enough for deployment requires careful judgment by the client health system.

**Box 5.2 Native EHR capabilities: Epic example**

Epic has released the *Genomic Indicators* module which gives genetic data a home in the EHR. Using the module, genetic data can be entered into the EHR in a structured format where it can then be used to trigger decision support and provide useful information to providers. Genomic indicators can also be accessed through the Epic patient portal where information can be delivered to patients based on their genetic markers. Several institutions have now implemented pharmacogenomics using Genomic Indicators (Caraballo et al., 2020; Gill et al., 2021; Mandel et al., 2016; Walton et al., 2020). Epic has adopted the legacy HL7 V2.5 genomic report standard for their underlying data architecture (HL7 Standards Product Brief - HL7 Version 2 Implementation Guide: Clinical Genomics, 2013). However, since this data structure does not lend itself well to pharmacogenomics, many implementations are not using this data architecture to store pharmacogenomic variants but rather use HL7 observation/result (OBX) segments to populate genomic indicators.

Genomic indicators have a well-established place in the Epic interface and serve as a new type of patient information similar to imaging results. CDS support can be launched based on genomic indicators using Epic's existing CDS architecture. This architecture allows for passive and active decision support and allows for dialogues that can suggest new medications and dosing with the ability to change the order through the alert dialogue.

Genomic indicators are present in the patient portal where patients can access information related to their genomic variants. Genomic indicators can be manually entered, or custom interfaces can be developed to import laboratory data and generate genomic indicators from it. At the time of writing, genomic indicators do not have support for Infobuttons, but there is the capability to access external resources through custom web links.

## Third-party clinical decision support systems and application programming interfaces

A significant challenge to implementing pharmacogenomics systems in the EHR in a large healthcare system is the need to rely on an internal EHR team that has many competing priorities within a healthcare ecosystem. One way around this is to build or buy an external application that interfaces with the EHR through an Application Programming Interface (API). An API allows for dynamic access to content which can be maintained by a third-party vendor. API interactions for CDS can open up streamlined workflows that benefit the clinician, caregiver, and patient. It is possible to send data about the patient context through an API so that the third-party service can provide context-informed (specific) guidance. Third-party CDS rules could potentially be more recent and comprehensive than what is available in the EHR vendor's CDS functionality. Third-party CDS may also provide more current recommendations than are allowed by an internal biomarker-based system due to the ability to connect in real time to data sources maintained by content experts. Key points and functionalities of some APIs, along with common abbreviations and terminology (Table 5.1) are reviewed in this section (McDonald et al., 2003).

## Fast Health Interoperability Resources

Fast Health Interoperability Resources (FHIR) is a standards framework for exchanging healthcare information developed by the HL7 community. FHIR is based on widely adopted web standards, including JSON and HTTP. It allows EHR systems to share data and communicate with other systems using REST web services, where REST stands for representational state transfer. RESTful interfaces are a key feature of system integration in the modern age of computing. They both simplify system interaction and, when combined with authentication protocols such as Open Authorization (OAuth) and secure data, transfer protocols such as SSL and Virtual Private Networks (VPNs), allow for patient information to travel securely outside of the EHR firewall.

There are several ways that FHIR can be used in pharmacogenomic CDS which are detailed in an implementation guide written by the HL7 Clinical Genomics working group (HL7 Standards Product Brief - HL7 Version 2 Implementation Guide: Clinical Genomics, 2013). However, adoption of these is not yet widespread and vendors have taken varying approaches to overcome this limitation. In the model used by YouScript (Invitae; San Francisco, CA), patient genomic data are sent separately to YouScript, whereas within the approach used by Coriell Life Sciences (Philadelphia, PA), patient data can pass through the API interaction but is also stored by Coriell.

## SMART on FHIR

As a standard way to interface with an EHR and transmit data, Health Level Seven (HL7) supports the use of Substitutable Medical Applications and Reusable Technologies (SMART) on FHIR (SMART on FHIR). This

**Table 5.1** Terminology for information exchange in clinical decision support.

| Term or abbreviation | Meaning |
|---|---|
| HL7 | Health Level 7 is a set of clinical standards and messaging formats that provide a framework for the management, integration, exchange, and retrieval of electronic information across different healthcare systems. HL7 standards are developed and maintained by Health Level Seven International, a healthcare standards organization. |
| LOINC | Logical Observation Identifiers Names and Codes (LOINC) is a database and universal standard for identifying medical laboratory observations. LOINC is mandated for use by the Office of the National Coordinator in the United States and is very commonly used internationally to provide a specific code to identify the type/kind of observation. For example, using a LOINC code can tell a CDS tool if a specific lab test is on a list of known pharmacogenomics tests. |
| RxNorm | RxNorm has normalized names for medical drugs and contains additional metadata from other medicine databases such as First Databank, Micromedex, and United States Pharmacopeia (USP). Additionally, formulation information is recorded and coded. The collected information and links to other medication databases roll-up to a single coded concept in RxNorm. Through use of a RxNorm code, a clinical decision support system is able to unambiguously identify a medication and can learn more about the medication. In pharmacogenomics, this can be used to identify medications for CDS. |
| USAN | United States Adopted Names (USAN) is a systematic standard of generic drug names. While US-based USAN works with the World Health Organization (WHO) to keep consistent with international naming conventions, there are differences between countries. USAN is consistent internally to the US. If used, ensure the CDS is using USAN as well as the local EHR. |
| SNOMEDCT | SNOMEDCT is a highly utilized and regarded standard for medical terms. In the arena of pharmacogenomic CDS, a list of the categorization of molecular phenotypes has been created in SNOMEDCT. |

Abbreviations for terms commonly used in pharmacogenomic CDS.

standard provides a reliable, consistent means for an EHR to execute a third-party application and for that application to interact with data stored in the EHR (Mandel et al., 2016). SMART apps are written as services that the EHR can launch within the clinical workflow using a URL. The app receives permission from the EHR to access data using FHIR queries. Once the EHR launches the app, the app's user interface is presented within the user interface of the EHR. This overcomes the limitation of third-party applications that operate outside of the EHR in that they require users to have multiple windows or devices in operation. Developers of SMART apps still need to work with the healthcare organization to set up access to the app which must be hosted in a HIPAA-compliant environment. Depending on the SMART app developer, either the developer's organization or the app hosting service needs to enter into a Business Associate Agreement for full HIPAA compliance.

## CDS Hooks

Whereas SMART on FHIR provides a standard way for an application to interface with the EHR, CDS Hooks allows EHR systems to trigger real-time CDS at specific steps within the clinical workflow (https://cds-hooks.hl7.org/). The CDS-Hooks project was launched by the SMART Team at Boston Children's Hospital, and is one of the SMART Health IT portfolio projects.

The specification is open source and has four basic components:

1. EHR—triggers the CDS Hook and processes the CDS service response.
2. Service—the system that receives a CDS request, applies CDS logic, and returns a response which can include information and action recommendations.
3. Hook—the clinical workflow step that triggers the event. The EHR can register different CDS services under each hook. Currently, there are three standard hooks (patient-view, order-select, and order-sign). Each CDS service is sent a resource containing the trigger event, patient information, and contextual information. Secure information on how the service can acquire more information is included in the resource and the service can use that information to obtain pertinent details such as genomic data repository. The amount of context sent to the service is usually kept to a minimum. For high-data consuming services, it is expected that the service will retrieve much of the context needed to calculate the response.

4. Card—a structured response from the service that provides recommendations such as alternative medications to order or pharmacogenomics clinical advice.

Recent publications provide detailed pharmacogenomics examples of the use of CDS Hooks (Dolin et al., 2018; Watkins & Eilbeck, n.d.). A basic example of CDS Hooks is shown in Fig. 5.2.

Briefly, a CDS Hooks interaction begins with a trigger event in the EHR. This trigger event can be as basic as opening a patient chart (patient-view), a medication order during computerized provider order entry (order-sign), or a display of the medication in a list of active medications (order-select). The EHR system then sends a message called a CDS Request to a CDS Service to which the EHR has subscribed. The CDS Request provides the context of the trigger event, a reference to the patient, and information on where/how the CDS service can obtain additional relevant information using FHIR queries. The CDS service can use that information to run CDS logic and arrive at a recommendation. The CDS service then returns the recommendation as a "card" resource to the EHR. The card has mechanisms to allow for a presentation of summary alert information, informative additional information, and further actions such as canceling an order or creating a new order. It is also possible to present supporting evidence and URL that the clinician could use trigger a SMART CDS application that might have more information to inform the clinician's final decision on how to handle the alert.

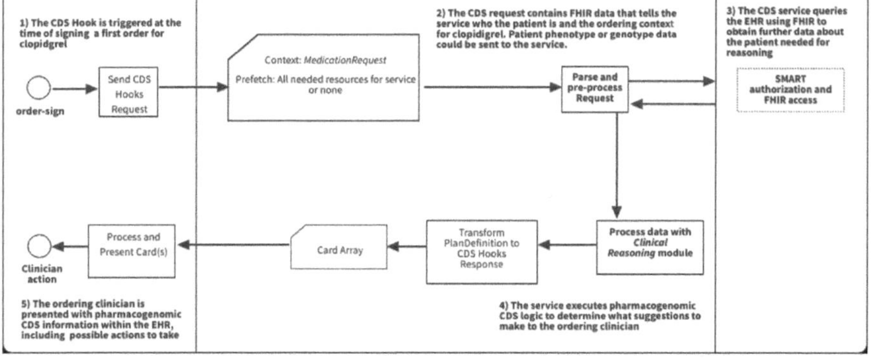


**Figure 5.2** A pharmacogenomic example of CDS Hooks. The figure shows five steps that would occur using CDS Hooks for pharmacogenomic CDS using a first-time order for clopidogrel as an example.

Similar to SMART apps, developers of CDS Hooks services need to work with the healthcare organization to set up access to the service which must be hosted in a HIPAA-compliant environment. Depending on the developer, either the developer's organization or the app hosting service needs to enter into a Business Associate Agreement for full HIPAA compliance.

### Infobuttons

Methodologies have been devised to deliver information to patients based on genomic data using Infobuttons as well as other proprietary technologies (Del Fiol et al., n.d.; Williams et al., 2018). Infobuttons is an HL7 standard for context sensitive content delivery that has been widely adopted by EHR vendors (Crump, 2018; Del Fiol, n.d.). Infobuttons are often visually placed next to terms on the problem list, medications, and laboratory tests. Clinicians can click on the Infobutton to link out to resources that are specific to the term they reside next to (Del Fiol et al., n.d.). PharmGKB currently serves as the sole HL7-compliant Infobutton pharmacogenomics data resource for targeted patient and provider information around pharmacogenomic variants (Barbarino et al., 2018). Other resources may be accessible through OpenInfobuttons, which allows for establishing connections to resources that are not compliant with the HL7 Infobutton standard (Heale et al., 2016).

## Storage and access of genomic data

The storing of genomic sequencing data for dynamic access through the EHR presents a significant and ongoing challenge, specifically whether to store genomic data within the native EHR or externally. The majority of EHR systems do not presently have adequate infrastructure to store genomic variants. Some capabilities do exist to translate genomic variants into genetic phenotypes that can, in turn, be stored in the EHR. However, this can present new challenges as knowledge changes and underlying variants have to be reclassified.

### Systems for communicating genomic data within the EHR

While the footprint of the genomic data for pharmacogenomics is very small as compared to the whole genome, many experts advocate storing entire genomic sequence via the EHR. This would enable providers or others to access specific data, such as pharmacogenomic information, polygenic risk scores, and germline pathogenic genetic variants, when necessary. One trending solution involves storing raw and intermediate genomic data externally to the EHR in a Genomic Archiving and Communication System (GACS) while housing only the reported data, or genetic phenotypes that are actively being used in patient care in the native EHR (Fig. 5.3) (A pharmacogenomics clinical decision support service based on FHIR and CDS Hooks. Methods of Information in Medicine, 2018; Starren et al., 2013). Such a system is somewhat analogous to existing Picture Archiving and Communication Systems (PACS) used in radiology. PACS could serve as a model on which to build new genetic data storage and access systems.

Standards for storing and transmitting genomic data are evolving, but even those that are starting to mature have had little adoption in the EHR and laboratory space. Maturation and adoption of standards and the development of standard APIs that allow ancillary genomics systems to interface with the EHR are needed to further scale pharmacogenomics. These are both evolving areas in genomics. Many research collaboratives are pushing for the adoption of FHIR standards for genomics (Devaney, 2018). GACS systems have been suggested to communicate with the EHR via SMART on FHIR interface (A pharmacogenomics clinical decision support service based on FHIR and CDS Hooks. Methods of Information in Medicine, 2018; Alterovitz et al., 2015). However, no production version EHR supports the FHIR genomics standards at the time of this publication.

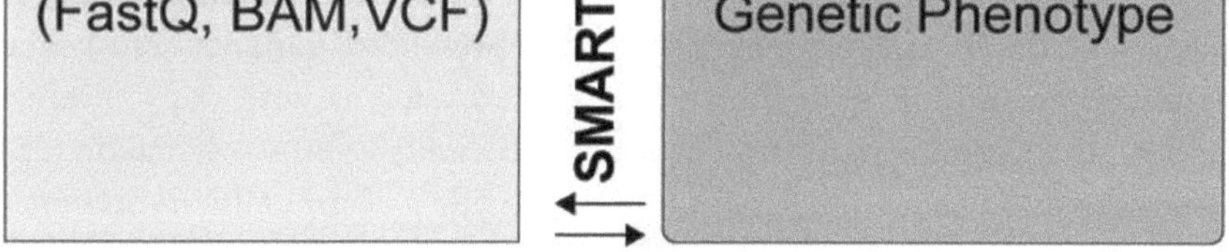


**Figure 5.3** Diagram of a Genomic Archiving and Communication System (GACS). In this method, raw and intermediate genomic data are stored externally to the EHR and only reported data, or genetic phenotypes that are actively being used in patient care, are housed within the native EHR.

While laboratories currently employ HL7 messaging to communicate laboratory results to Laboratory Information Systems (LIS), which are then relayed into the EHR, laboratories do not currently utilize the HL7 Genomic Laboratory Report format to transmit these data and most do not transfer any structured genetic data. The LIS is designed to facilitate the ordering and return of laboratory tests; however, most are not designed to store and transmit discrete genetic data. This has led some to question the role of the LIS in passing genetic data to the EHR as it adds a layer of complexity to the solution if an EHR can handle structured genetic data and the LIS does not have this capability. Because discrete genomic data residing in the LIS cannot be used in clinical care, one method to getting structured genetic data into the EHR may involve maintaining the existing dataflows for genetic data in the LIS and enabling a separate dataflow from the laboratory directly to the EHR as shown in *Box A* of Fig. 5.4. Within this figure, *Box B* demonstrates the flow of data into an LIS that is able to store and pass structured genetic data to the EHR, while *Box C* demonstrates a workflow with a GACS system implemented. Most third-party tools use a model similar to *Box C*, with the exception that the GACS only stores pharmacogenomic data and structured genetic data are accessed through API calls. Some have suggested that a modified version of the LIS might function as GACS system as shown in *Box D* of this figure.

Regardless of the strategy used to store and transmit pharmacogenomic data, challenges remain in determining the format that the data should be stored in. There is currently not a single universally accepted standard or format for molecular sequence data. Some have proposed that for simple variants, use of the Human Genome Variation Society (HGVS) normalization such as has been done with the NCBI's SPDI data model would be useful (Holmes et al., 2020). The basic model is the position of a change within a reference sequence, the reference sequence, and the molecular sequence change. NCBI's SPDI uses HGVS syntax to provide details to fill in the model for simple variants. Many pharmacogenomic star alleles can be modeled as the co-occurrence of SPDI represented variants. However, when chromosome level positions are affected, such as with copy number variation, the simple variant model is not sufficient. Even so, syntactic representations and the simple variant model provide a computable representation of many variants relevant to pharmacogenomics (Holmes et al., 2020).

For those using biomarkers, a standard for the molecular phenotype is required to harmonize between laboratory reports and CDS. The Clinical Pharmacogenetics Implementation Consortium (CPIC) offers several

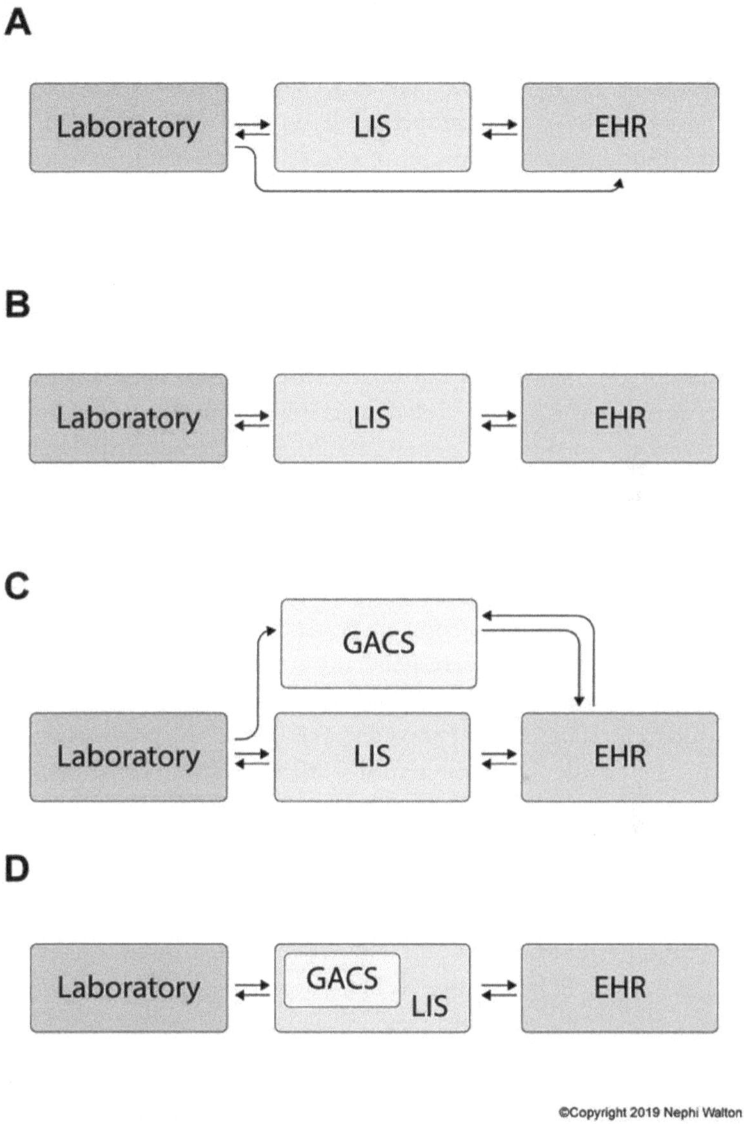


**Figure 5.4** Diagram of different genomic data flows from the laboratory to the EHR. Different methods of storing and communicating pharmacogenomics data to CDS that have been proposed. See the text of this chapter for more information.

molecular phenotype lists that could be considered definitive. However, if the reporting laboratory sends an interpreted molecular phenotype that does not match those lists, further mapping will be necessary. Human intervention has been one means of ensuring a match between the semantics used for biomarkers by a CDS vendor and the incoming results from genetic testing (Pasternak et al., 2020). Use of LOINC answer lists and specific LOINC codes has also been proposed. It must also be determined as to whether the biomarker is the presence of a star allele or molecular phenotype. See the Chapter "Scientific Evidence and Sources of Knowledge for Pharmacogenomics" in this book for a more detailed discussion of molecular phenotype standards and CPIC.

## Static versus dynamic reporting of pharmacogenomic data

Due to the dynamic nature of the pharmacogenomics field, pharmacogenomics reporting has become a challenge when working with traditional static PDF reports. In the past, genetic results have traditionally been issued through a printed or PDF report. An advantage of this method is that, when the need to update the clinical interpretation of a genetic result arose, it could be managed by amending and reissuing a new report since most variants are rare and a change would not affect a large number of cases. In addition, because clinicians specializing in genetics were the primary users of these static reports, changes were generally understood and interpreted by specialist clinicians for treatment purposes.

Unfortunately, static PDF reports are a challenge for pharmacogenomic precision medicine due to a combination of three factors: (1) the use of pharmacogenomic testing in large populations, with thousands of individuals having the same variants; (2) the large number of highly polymorphic variants evaluated; and (3) a rapidly evolving field that is constantly generating new information, establishing new genotype-phenotype correlations, and updating clinical recommendations. This has created a dilemma for institutions utilizing static PDF reports. When there are changes to genotype–phenotype correlations or clinical recommendations, potentially hundreds to thousands of patients could be affected (Box 5.3). At this scale, it is not feasible for new printed or PDF reports to be generated for every case. Even if an institution is able to generate new reports, there is still a concern as to which report would be used in the clinical decision-making process. Patients might not be aware of the new report or might have moved out of the given healthcare system taking with them the original static report. In the case of the latter, the new provider may not understand how to utilize current pharmacogenomic resources and thus may rely on out-of-date information when providing care.

**Box 5.3 Limitations of static reporting for pharmacogenomic test results**

With the intent of standardizing star allele classifications, in October 2019 the Clinical Pharmacogenetics Implementation Consortium (CPIC) updated its guidelines for *CYP2D6*. Revised guidelines indicated that a *1/*4 diplotype would no longer be classified as a normal metabolizer, but instead would be classified as an intermediate metabolizer (Caudle et al., 2020). The *4 allele is based on the c.506-1G > A variant which has an allele frequency of 0.138 in the normal population according to gnomAD (Gudmundsson et al., 2021). This could mean that up to 28% or more of individuals who had previously undergone *CYP2D6* testing would need an updated phenotype classification.

In this case, new reports would need to be issued for a portion of patients who had undergone *CYP2D6* testing in institutions that rely on static PDF reports. As an example, an organization that had tested 10,000 individuals and provided static reports would then need to reissue approximately 2800 updated PDF reports. This process is time-consuming, likely to increase in its impact as emerging discoveries are made, and likely not feasible. Alternatively, the institution must make the decision to leave the reports unrevised and rely on clinicians to understand and correctly interpret the genotype in light of revised guidelines. This real-world example highlights the need to move to more dynamic solutions for reporting pharmacogenomic test results that can be updated quickly across thousands of cases, such as an online secure report tool.

Many laboratories are moving to a more dynamic reporting method in which patients and providers can log in to a secure online system to view the results and see interpretations and recommendations for drug dosing. These reporting databases can be updated rapidly for all patients so that anyone logging in will see the most up-to-date interpretations for any variants in the patient's chart. If there is an update that could affect drug recommendations for a patient, the electronic system allows for patients and providers to be emailed or alerted to the changes. These reporting methods can be developed to meet College of American Pathology (CAP) and Clinical Laboratory Improvement Amendments (CLIA) requirements for reporting. It is important to consider different options for reporting and how this might impact clinical support. Currently there is no EHR solution that authors are aware of that supports this type of reporting.

## Cloud computing and clinical decision support

Cloud computing involves the scalable distribution of computer system resources over several remote services that can be accessed through the internet or an intranet. In general, these servers are managed by third-party vendors such as Amazon, Microsoft, and Google. These resources allow for rapid scaling of compute and storage needs and are well-suited for large datasets and computationally intensive operations that can be made faster by distributing them over a large number of processors. These systems have advantages in general related to scalability, ease of management, speed of access, and redundancy (in case a system goes down). While cloud computing is an ideal environment for managing and processing large genomic datasets such as whole exome sequencing and whole genome sequencing due to the data size and complexity, pharmacogenomics in its current state does not necessitate the use of cloud computing as it consists of a few genetic data points extracted from the whole genome. The storage of pharmacogenomic markers is a small fraction of what is required for a genome (<1/1,000,000). Data storage needs for pharmacogenomics will increase over time especially as pharmacogenomics biomarkers are combined with polygenic risk scores and other genetic data to better tailor care. While cloud computing has many advantages, it can be expensive and there have been concerns about security. The decision to deploy on a cloud-based system versus a local server does not necessarily affect the performance of the application and is more a question of system architecture and security than of pharmacogenomics CDS strategy.

## Conclusion

Many challenges exist in the implementation of pharmacogenomics CDS systems. Setting aside questions of standardization, an essential hurdle is finding a home for genomic data in the EHR. Several strategies have been developed in regards to storing and displaying pharmacogenomic data, such as adding this information to the problem list or to allergies. Neither choice is truly a home for genomic data, nor are they scalable solutions as the amount of clinical genomic data grows over time. As genomic data become more ubiquitous in patient care, it is important that there be a commonly accepted location for the data in the EHR, similar to imaging or other laboratory data.

As a part of this development, healthcare providers should be more educated about pharmacogenomics and other forms of genomic data and the role that they play in patient care. It is especially important because clinically relevant genomic data can influence care decisions throughout a patient's entire life. Currently, few patients have pharmacogenomic data present in the EHR and providers are unlikely to seek out these data in the EHR. Once an established "home" exists in the EHR for genomic data and providers are actively seeking it, then it will help alleviate the need for active decision support that is more interruptive to clinical workflow.

Maintaining CDS for pharmacogenomics over time is challenging, especially as the field grows. Moreover, the implementation of these guidelines is often partially dictated by clinical preferences rather than hard and fast guidelines and is certainly not uniform across institutions. This lack of uniformity is part of the challenge related to adoption of different pharmacogenomic platforms. There is disagreement on how to implement PGx CDS among institutions and often even amongst individual providers in the same institution. While dynamic reporting may help institutions keep up with changes in guidelines, current EHR infrastructures do not generally support this type of reporting.

An additional challenge is how to handle the longitudinal nature of genetic data, with germline genomic data continuing to influence a patient's health for their entire lifetime. Currently, genetic results are not accessible in a centralized manner but drift between multiple test results and, in some cases, miscellaneous test results. Programmatic and clinician access are different problems to address, but both require a centralized collection of genomic data.

Many EHR systems were originally constructed to allow for manual input of pharmacogenomic data because standards were not widely available or adopted at the time of implementation. The manual entry of genetic data is not feasible in a clinical environment and the automated transfer through a standard interface is necessary in order to facilitate structured data in the EHR. Creating separate custom interfaces for every laboratory from which institutions receive pharmacogenomic data is not scalable and constrains healthcare systems to specific relationships with pharmacogenomic testing providers and lessens patient/provider choice when selecting pharmacogenomics testing options. The FHIR and CDS Hooks standards have matured to a level that does allow for effective capture and transmission of pharmacogenomic data. Both are likely to have an important role in the future of pharmacogenomic CDS as adoption among laboratories and EHR systems grows.

## References

Alterovitz, G., Warner, J., Zhang, P., Chen, Y., Ullman-Cullere, M., Kreda, D., & Kohane, I. S. (2015). SMART on FHIR Genomics: Facilitating standardized clinico-genomic apps. *Journal of the American Medical Informatics Association, 22*(6), 1173–1178. https://doi.org/10.1093/jamia/ocv045

Barbarino, J. M., Whirl-Carrillo, M., Altman, R. B., & Klein, T. E. (2018). PharmGKB: A worldwide resource for pharmacogenomic information. *Wiley Interdisciplinary Reviews: Systems Biology and Medicine, 10*(4), e1417. https://doi.org/10.1002/wsbm.1417

Bell, G. C., Crews, K. R., Wilkinson, M. R., Haidar, C. E., Hicks, J. K., Baker, D. K., Kornegay, N. M., Yang, W., Cross, S. J., Howard, S. C., Freimuth, R. R., Evans, W. E., Broeckel, U., Relling, M. V., & Hoffman, J. M. (2014). Development and use of active clinical decision support for preemptive pharmacogenomics. *Journal of the American Medical Informatics Association, 21*(2), e93–e99. https://doi.org/10.1136/amiajnl-2013-001993

Caraballo, P. J., Sutton, J. A., Giri, J., Wright, J. A., Nicholson, W. T., Kullo, I. J., Parkulo, M. A., Bielinski, S. J., & Moyer, A. M. (2020). Integrating pharmacogenomics into the electronic health record by implementing genomic indicators. *Journal of the American Medical Informatics Association, 27*(1), 154–158. https://doi.org/10.1093/jamia/ocz177

Caudle, K. E., Sangkuhl, K., Whirl-Carrillo, M., Swen, J. J., Haidar, C. E., Klein, T. E., Gammal, R. S., Relling, M. V., Scott, S. A., Hertz, D. L., Guchelaar, H.-J., & Gaedigk, A. (2020). Standardizing CYP2D6 genotype to phenotype translation: Consensus recommendations from the clinical pharmacogenetics implementation Consortium and Dutch Pharmacogenetics Working Group. *Clinical and Translational Science, 13*(1), 116–124. https://doi.org/10.1111/cts.12692

Collins, T., Power, K., McCallie, D., & Owings, R. (2019). Finding a place for pharmacogenetics in the electronic health record. *Clinical Pharmacology and Therapeutics, 106*(2), 295–297. https://doi.org/10.1002/cpt.1504

Del Fiol. (n.d.). Integrating genetic information resources with an EHR. In, *AMIA annual symposium proceedings 2006*: American Medical Informatics Association.

Del Fiol, Rocha, R. A., & Clayton, P. D. (n.d.). Infobuttons at intermountain healthcare: Utilization and infrastructure. In, *AMIA annual symposium proceedings. 2006.* American Medical Informatics Association.

Crump, J. K. (2018). Prototype of a standards-based EHR and genetic test reporting tool coupled with HL7-compliant infobuttons. *AMIA Summits on Translational Science Proceedings 2018*, 330.

Devaney, S. (2018). *All of us research program.*

Dolin, R. H., Boxwala, A., & Shalaby, J. (2018). A pharmacogenomics clinical decision support service based on FHIR and CDS hooks. *Methods of Information in Medicine, 57*(4), E115–E123. https://doi.org/10.1055/s-0038-1676466

Gill, P., Yu, F., Porter-Gill, P., Boyanton, B., Allen, J., Farrar, J., Veerapandiyan, A., Prodhan, P., Bielamowicz, K., Sellars, E., Burrow, A., Kennedy, J., Clothier, J., Becton, D., Rule, D., & Schaefer, G. (2021). Implementing pharmacogenomics testing: Single center experience at Arkansas Children's Hospital. *Journal of Personalized Medicine, 11*(5), 394. https://doi.org/10.3390/jpm11050394

Gudmundsson, S., Singer-Berk, M., Watts, N. A., Phu, W., Goodrich, J. K., Solomonson, M., Rehm, H. L., MacArthur, D. G., O'Donnell-Luria, A., & Genome Aggregation Database (gnomAD) Consortium. (2021). Variant interpretation using population databases: Lessons from gnomAD. *Human Mutation*, 1–19. https://doi.org/10.1002/humu.24309

Heale, B. S. E., Overby, C. L., Del Fiol, G., Rubinstein, W. S., Maglott, D. R., Nelson, T. H., Milosavljevic, A., Martin, C. L., Goehringer, S. R., Freimuth, R. R., & Williams, M. S. (2016). Integrating genomic resources with electronic health records using the HL7 infobutton standard. *Applied Clinical Informatics, 7*(3), 817–831. https://doi.org/10.4338/ACI-2016-04-RA-0058

Hinderer, M., Boeker, M., Wagner, S. A., Lablans, M., Newe, S., Hülsemann, J. L., Neumaier, M., Binder, H., Renz, H., Acker, T., Prokosch, H. U., & Sedlmayr, M. (2017). Integrating clinical decision support systems for pharmacogenomic testing into clinical routine- a scoping review of designs of user-system interactions in recent system development. *BMC Medical Informatics and Decision Making, 17*(1). https://doi.org/10.1186/s12911-017-0480-y

*HL7 standards product brief - HL7 version 2 implementation guide: Clinical genomics*.(2013). http://www.hl7.org/implement/standards/product_brief.cfm?product_id=23.

Holmes, J. B., Moyer, E., Phan, L., Maglott, D., & Kattman, B. (2020). SPDI: Data model for variants and applications at NCBI. *Bioinformatics (Oxford, England), 36*(6), 1902–1907. https://doi.org/10.1093/bioinformatics/btz856

Keller, J. P. (2012). Clinical alarm hazards: A top ten health technology safety concern. *Journal of Electrocardiology, 45*(6), 588–591. https://doi.org/10.1016/j.jelectrocard.2012.08.050

Kizzier-Carnahan, V., Artis, K. A., Mohan, V., & Gold, J. A. (2019). Frequency of passive EHR alerts in the ICU: Another form of alert fatigue? *Journal of Patient Safety, 15*(3), 246–250. https://doi.org/10.1097/PTS.0000000000000270

Mandel, J. C., Kreda, D. A., Mandl, K. D., Kohane, I. S., & Ramoni, R. B. (2016). SMART on FHIR: A standards-based, interoperable apps platform for electronic health records. *Journal of the American Medical Informatics Association, 23*(5), 899–908. https://doi.org/10.1093/jamia/ocv189

McDonald, C. J., Huff, S. M., Suico, J. G., Hill, G., Leavelle, D., Aller, R., Forrey, A., Mercer, K., DeMoor, G., Hook, J., Williams, W., Case, J., & Maloney, P. (2003). LOINC, a universal standard for identifying laboratory observations: A 5-year update. *Clinical Chemistry, 49*(4), 624–633. https://doi.org/10.1373/49.4.624

Pasternak, A. L., Ward, K. M., Ateya, M. B., Choe, H. M., Thompson, A. N., Clark, J. S., & Ellingrod, V. (2020). Establishment of a pharmacogenetics service focused on optimizing existing pharmacogenetic testing at a large academic health center. *Journal of Personalized Medicine, 10*(4), 1–10. https://doi.org/10.3390/jpm10040154

A pharmacogenomics clinical decision support service based on FHIR and CDS hooks. *Methods of Information in Medicine, 57*, (2018).

Starren, J., Williams, M. S., & Bottinger, E. P. (2013). Crossing the omic chasm: A time for omic ancillary systems. *JAMA - Journal of the American Medical Association, 309*(12), 1237–1238. https://doi.org/10.1001/jama.2013.1579

Walton, N. A., Johnson, D. K., Person, T. N., Reynolds, J. C., & Williams, M. S. (2020). Pilot implementation of clinical genomic data into the native electronic health record: Challenges of scalability. *ACI Open, 4*(2), e162–e166. https://doi.org/10.1055/s-0040-1721490

Watkins, M., & Eilbeck, K. (n.d.). FHIR lab reports: Using SMART on FHIR and CDS hooks to increase the clinical utility of pharmacogenomic laboratory test results. *AMIA Joint Summits on Translational Science Proceedings*. AMIA Joint Summits on Translational Science, 2020, 683–692.

Williams, J. L., Rahm, A. K., Zallen, D. T., Stuckey, H., Fultz, K., Fan, A. L., Bonhag, M., Feldman, L., Segal, M. M., & Williams, M. S. (2018). Impact of a patient-facing enhanced genomic results report to improve understanding, engagement, and communication. *Journal of Genetic Counseling, 27*(2), 358–369. https://doi.org/10.1007/s10897-017-0176-6

Williams, M. S., Taylor, C. O., Walton, N. A., Goehringer, S. R., Aronson, S., Freimuth, R. R., Rasmussen, L. V., Hall, E. S., Prows, C. A., Chung, W. K., Fedotov, A., Nestor, J., Weng, C., Rowley, R. K., Wiesner, G. L., Jarvik, G. P., & Fiol, G. D. (2019). Genomic information for clinicians in the electronic health record: Lessons learned from the clinical genome resource project and the electronic medical records and genomics network. *Frontiers in Genetics, 10*. https://doi.org/10.3389/fgene.2019.01059

CHAPTER SIX

# The role of machine learning to predict treatment response and target drug therapies

**Casey Overby Taylor[1,2,3], James M. Stevenson[4]**
[1]Department of Medicine, Division of General Internal Medicine, Johns Hopkins University School of Medicine, Baltimore, MD, United States
[2]Department of Biomedical Engineering, Johns Hopkins University, Baltimore, MD, United States
[3]Institute for Computational Medicine, Johns Hopkins University, Baltimore, MD, United States
[4]Department of Medicine, Division of Clinical Pharmacology, Johns Hopkins University School of Medicine, Baltimore, MD, United States

## Contents



## Introduction

While progress applying the growing understanding of pharmacogenes in precision medicine continues, multiple factors should be considered when aiming to maximize drug efficacy and reduce the risk of side effects and adverse drug reactions. Strategies in artificial intelligence (AI) and machine learning (ML) are promising to bring together such factors to discover patient subtypes and to predict treatment response for those subtypes. This

*Clinical Decision Support for Pharmacogenomic Precision Medicine*
ISBN: 978-0-12-824453-1
https://doi.org/10.1016/B978-0-12-824453-1.00002-6

chapter explores such strategies and the potential for machine learning-driven clinical decision support to deliver drug treatment guidance targeted to certain patient subtypes.

## P4 medicine and deep phenotyping to target drug therapies

Precision medicine is a model of clinical care that is predictive, personalized, preventive, and participatory ("P4 Medicine") (Hood & Flores, 2012; Hood & Friend, 2011). P4 medicine recognizes that a unique combination of genetics, lifestyle, environmental factors, and health history contributes to an individual's physiological state. In addition, technological advances during the past decade have contributed to current capabilities to capture a variety of signals contributing to physiological state. By deeply characterizing the state of disease (i.e., *deep phenotyping*), patients can be stratified into subtypes to facilitate personalizing and targeting drug therapies. Deep phenotyping strives to comprehensively collect, aggregate, and analyze data from an individual to be able to ultimately suggest personalized treatments and strategies to improve disease trajectories (Paquette et al., 2020; Yurkovich et al., 2020).

However, translational barriers remain to connect data from the different settings where genomic, environmental, and clinical data are collected (Kim et al., 2021). One key source for clinical data is the electronic health record (EHR), where there is potential to bring these data worlds together with a growing number of EHR systems that link clinical records of patients with their genetic data (Cho et al., 2012; Gottesman et al., 2013; Kohane, 2011; Sankar & Parker, 2013; Vimal et al., 2021). Despite successes in identifying of genetic signatures with high predictive power to divide patients into treatment groups, the phenotypes and outcomes used to train models may lack the heterogeneity needed for clinical application. Identifying phenotypes and outcomes in EHRs represents a significant informatics challenge due to heterogeneity, incompleteness, and the dynamic nature of EHR data (Weiskopf & Weng, 2013).

For instance, disease severity scales used to measure baseline symptoms and response to treatment in research studies are rarely used in clinical settings (Uher, 2016). Even if researchers successfully use ML on research data to predict medication response, applying these approaches to clinical settings will be difficult given the lack of phenotypes and outcomes typically recorded in clinical practice. By way of example, researchers recently used ML methods to predict treatment response to the antidepressants citalopram and escitalopram using data from three research studies (Athreya et al.,

2019). The predictive models utilized genetic data and baseline symptom rating scales (e.g., Quick Inventory of Depressive Symptomatology or Hamilton Depression Rating Scale) as predictors for the treatment response. The predictors in the model—genetic information and a quantitative representation of depression severity—would not typically be accessible in most clinical settings. Going forward, this translational barrier may be addressed by wider clinical adoption of measurement-based care and genomic medicine (Guo et al., 2015; Uher, 2016).

Even at present, EHRs remain a promising source for deep phenotyping. EHR systems provide a wider array of phenotypes when compared with traditional population-based studies (Casey et al., 2016). EHR-derived patient populations are also representative of local healthcare system populations. Deep phenotyping techniques may allow us to discover so-called *verotypes* (Boland et al., 2013), or patient subgroups with shared genotype, phenotype, and clinical disease subtype. It may then be possible to select tailored treatment options for future patients matching a particular verotype. AI and ML methods are well-suited to help address translational barriers to discover patient subgroups and to predict outcomes with different treatment options for those subgroups.

## Machine learning to identify groups and predict treatment response

The phrase *artificial intelligence* (AI) refers to a subfield of computer science that studies computer applications that can perform tasks that normally require human intelligence. *Machine learning* (ML) is a technique for achieving AI that can be further divided into *unsupervised* or *supervised* methods. Unsupervised ML approaches are applied to uncover patterns in the data, where one application is to infer subtypes (or verotypes) of a disease. Supervised approaches aim to build models that are predictive of observed phenotypes such as drug response (Fig. 6.1).

Emerging AI and ML capabilities can be exploited to convert deep phenotyping data into actionable drug therapy possibilities. Indeed, the computational capacity to analyze multiple types of omics data, including pharmacogenomic data, already exists (e.g., NHGRI's Genomic Data Science Analysis, Visualization, and Informatics Lab-space, AnVIL (Schatz et al., 2022)). In addition, some ML models to predict responsiveness to drug therapy have already shown promise (Guan et al., 2019; Ke et al., 2010; Sieberts et al., 2016).

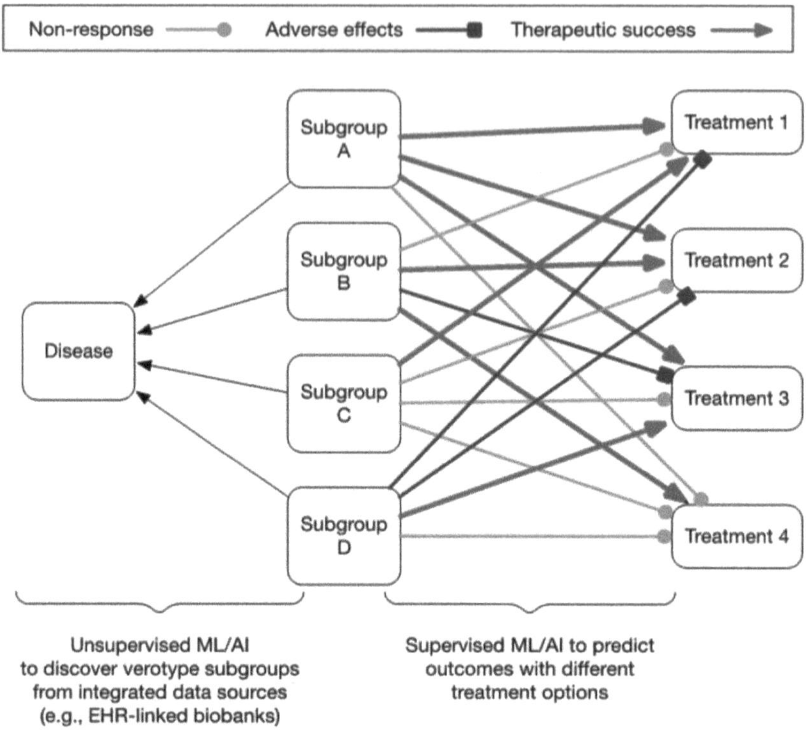


**Figure 6.1** Machine learning and artificial intelligence to predict treatment outcomes. Machine learning may be used with supervised or unsupervised methods to uncover patterns in data or build predictive models.

## Data preparation and data to inform clinical decision support

Frameworks currently used in ML and systems genomics for the integrative analysis of multiomics data can be helpful in building an ML model from multiple sources (Li et al., 2017; Lin & Lane, 2017). Additional data preparation, however, is likely needed to extract data from an institution's EHR system and to transform it into a format that can be used by these frameworks. A summary of recommendations provided by Bates et al., to address EHR data preparation needs, includes determining which variables will be taken from standardized sources versus unstructured data or free text, making explicit data assumptions regarding missing data and censoring, and defining how the study sample was obtained and the extent to which it was socio-demographically diverse. In addition, the model building process should involve clinician input to understand issues around documentation biases (Bates et al., 2020). Practical approaches and timing on how to do this, however, have yet to be determined.

Structured data required for accurate clinical decision support (CDS) should be available in the EHR to apply an ML model clinically. A major cause for failed CDS more broadly is the mismatch between information needed for decision rules and the availability of that information in the EHR (Kawamoto & McDonald, 2020). Given that EHR data are collected for routine clinical care or billing (Goroll, 2017), it can be difficult to extract valid condition or treatment response. For example, decision rules about warfarin dosing ideally utilize information about age, race, smoking status, concomitant medications, and genetics (Gage et al., 2008). The relevant patient data may be distributed across data sources to which the EHR does not have access. Recommendations made by Kawamoto et al., to address this challenge include (Kawamoto & McDonald, 2020)

1. Verification that the EHR includes the data needed by candidate rules, and if the data are inadequate, surrogates may be used or data can be collected via staff or patients;
2. Natural language processing (Demner-Fushman et al., 2009) can be used to access content of free-text notes for CDS, but it is still uncommon in this context; and
3. External data sources can also be accessed.

Emerging data standards such as the Fast Healthcare Interoperability Resources (FHIR) from Health Level Seven (HL7, 2021) and the adoption of the US Office of the National Coordinator for Health Information Technology (ONC, 2019) proposed rules requiring certified healthcare information systems to support FHIR and associated data coding standards may help to enable more comprehensive data from external sources for CDS.

## Validation and evaluation of machine learning–driven CDS

External validation and bias correction are needed to build an ML model (Bates et al., 2020). Indeed, a model trained from one population may not apply to another population, which has implications for clinical utility and the applicability of predictions for different contexts (Rajkomar et al., 2019). Validation recommendations made by Bates et al. include to

1. validate performance in a separate data set that is ideally from another institution;

2. learn from existing models that have been developed to predict the same outcome; and
3. measure accuracy by using multiple segments of a dataset (Bates et al., 2020).

Recommendations for initiating CDS more broadly for ML-based CDS rules can also be applied. For example, Kawamoto et al. recommend examining CDS results against patient data or conducting formative evaluations prior to initiating CDS (Kawamoto & McDonald, 2020). Any observed errors should then be corrected prior to implementation. For example, consider a framework used to evaluate EHR-based phenotype algorithms to validate that information needed for decision rules are in the EHR (Overby et al., 2013). In this context, the purpose would be to (a) detect any issues with measurement approaches used by the decision rules (e.g., Is the strategy to measure the presence of a condition robust to missing data from some fields?), and (b) determine the performance of the decision rules (e.g., positive predictive value, negative predictive value, sensitivity, and specificity).

## Explanation and interpretation of machine learning–driven CDS

A well-known challenge to the clinical use of ML models is that the output can be difficult to interpret (Rajkomar et al., 2019). Building trust is a requirement for the clinical adoption of such models. For this reason, the transparency of a model underlying CDS for treatment guidance is essential. Thus, it is unsurprising that the majority of endorsements for the clinical use of pharmacogenomics made by professional organizations are for gene–drug interactions. It is relatively straightforward for gene–drug interaction-based CDS to be implemented with rules, a model type recognized as interpretable. For other so-called "black box" models as with deep learning, however, there are additional challenges to their adoption. First, explaining the functionality of and rationale for recommendations made by complex computational models is a technically challenging problem. Furthermore, explanations may provide little meaningful information to a clinical user, which raises the question of their value. Second, disclosing computational model details may compromise the privacy of data subjects (Vayena et al., 2018) or may not be a preferred business choice for proprietary CDS systems. Despite these challenges, computational models for genomics and pharmacogenomics are becoming more abundant, and transparency is needed to build trust (Vayena et al., 2018) between CDS system implementers and the clinical users of the CDS for pharmacogenomics. Importantly, explanation of CDS-delivered recommendations does not hinge on a clinical user's understanding how the model works.

Given the highlighted challenges to explaining model functionality, this section focuses on explanations according to specific clinical actions they are intended to support. *Counterfactual explanations* (Mittelstadt et al., 2019; Mothilal et al., 2020) provide one approach to do this by showing "what-if" explanations for model output. The work of others to explore the purpose of counterfactual explanations for understanding, contesting, and altering model-driven decision-making (Wachter et al., 2017) provides a useful framework to explore explanations for model-driven CDS for pharmacogenomics. In particular, explanations can assist clinicians with (a) understanding why a particular recommendation was made, (b) providing reasons to decline a recommendation, and (c) understanding what behaviors patients can change to reach a better outcome in the future.

Counterfactual explanations have been described as statements with the form: "Score $p$ was returned because variables $V$ had values ($v1$, $v2$, ...) associated with them. If $V$ instead had values ($v1'$, $v2'$, ...) and all other variables had remained constant, score p' would have been returned." An example scenario might be from a study indicating that alcohol misuse was associated with major bleeding on warfarin, with potentially higher risks among patients on long-term warfarin and with *CYP2C9* and/or *VKORC1* genetic variants (Roth et al., 2015). An example CDS recommendation for pharmacogenomics would be to apply a classifier that returns a risk for a bleeding event as a score. Counterfactuals could be generated to answer the question "What would have to be different for this individual to have a bleeding event risk score of 0.2?" In response to this question, multiple diverse counterfactual explanations should be provided to a clinician to allow for changes for a more favorable future outcome.

## Understanding a recommendation

To understand a CDS recommendation for pharmacogenomic testing, counterfactuals can provide insight into reasons that led to that recommendation. Simple "if-then" statements such as those in the prior example provide transparent and easily accessible explanations of how following a particular recommendation might affect the patient. The influence of only a minimal number of patient facts and variables on a recommendation can help a clinician make a responsible decision and can avoid disclosure of the internal logic of complex models that can be challenging for nonexperts to understand.

## Recommendation decline reasons

To help clinicians identify a need to decline a recommendation due to inaccurate or incomplete data, counterfactual explanations can provide the reasons that led to a recommendation. Doing so enables a clinician to assess whether the reasons are legitimate and can provide reasoning to decline a recommendation. For example, if documentation of a CYP2D6 poor metabolizer phenotype and a prescription of a CYP2D6 substrate led to a recommendation not to prescribe yet another CYP2D6 substrate, a clinician may learn from the patient that they are no longer taking the other CYP2D6 substrate. This discovery would then provide grounds to decline the recommendation.

## Enabling behavior change messaging

To help clinicians communicate to patients what changes in behavior may impact an outcome underlying the recommendation, counterfactual explanations can provide information about the patient factors and key variables that would result in a different recommendation. Doing so can provide clinicians with clear messaging for how behaviors can be altered to receive a different result in the future. For example, consider a counterfactual explanation of a recommendation to reduce alcohol consumption based on results from a patients' screening for alcohol misuse and the presence of genetic variants involved in warfarin metabolism and anticoagulation mechanisms. An explanation could indicate how risk for bleeding events change with changes in alcohol consumption.

Interpretable computational modeling is an active area of research and use of counterfactual explanations is only one of many strategies. ML feature attribution methods that assign "weights" to features used to make predictions is another strategy to provide data interpretability (Lundberg et al., n.d.; Ribeiro et al., 2018). The simplicity of counterfactual explanations does have limitations and is not appropriate in all scenarios. For example, they do not provide statistical evidence to assess models for fairness or racial bias. They do, however, provide a first step toward transparency while also overcoming the challenge of preserving privacy since no individual data from a training dataset or detailed information about the model need to be disclosed. Such transparency would potentially increase trust and downstream adoption of CDS for pharmacogenomics.

## Implementation and dissemination of machine learning–driven CDS

Standard strategies used to introduce CDS into healthcare can inform implementation of ML-driven CDS in a way that is most useful to clinicians (Tcheng et al., 2017). Although ML-based guidance is translated into rules to drive CDS (e.g., membership to subgroup A, a medication order), some uncertainty will remain when applying ML insights in clinic (Topol, 2019). To help address the issue of uncertainty, some of the following strategies described by Kawamoto et al. can be particularly helpful (Kawamoto & McDonald, 2020). First, choose decision rules that are consistent with local care process. For example, previous pharmacogenomics-based CDS is often implemented at the time of a medication order. This timing, however, may not be ideal due to it being at the end of the decision-making process. An alternative strategy for a decision rule might be to determine membership to a specific subgroup or verotype group to drive CDS. Second, select meaningful evaluation measures such as compliance with standards of care (e.g., order completion time), cost (e.g., per patient), and intermediate outcomes (e.g., presence of side effects). Given that many models involve retrospective back-testing of algorithms on existing data, it is particularly important to prospectively monitor and study ML-guided decisions support. Last, consider how to deploy and study the CDS beyond the original host organization. CDS should be described in sufficient detail to enable others to replicate its deployment. For ML-guided decision support, reporting may also need to involve steps to build the model with local data to achieve optimal performance.

## Polygenic risk profiling

To date, most pharmacogenomics CDS implementations provide drug therapy guidance for drug–gene pairs. More complex data analyses to discover patient subtypes and to predict treatment response, however, are on the rise. Thus, there will be more opportunities for new drivers of CDS from those analyses. While the focus of this chapter has been on ML-guided decision support to target drug therapy, there are also recent studies demonstrating the utility of polygenic risk profiling.

Polygenic risk profiling on the basis of calculated polygenic risk scores (PRS) can be used to prioritize therapeutic interventions for certain patient subtypes (Torkamani et al., 2018). A recent review of the potential for PRS to target drug therapy, however, has highlighted a need to better stratify patient subgroups and target those predicted to have the highest therapeutic response rates (Gibson & Barsh, 2019). For example, many studies show clinical utility for high-risk populations where the response rates are often higher (e.g., PRS for statin response).

One recommendation is to evaluate therapeutic response rates across the distribution of PRS scores (Gibson & Barsh, 2019). As suggested in this chapter, there is also an opportunity to take a stepwise approach to discover patient subtypes prior to building a prediction (ML or otherwise) around therapeutic response. Phenotype profiling is a promising strategy to define patient subtypes and also to bring in additional sources for deep phenotyping including EHR (Linder et al., 2021) and wearables. Demonstrating the promise of phenotype profiles for genomics applications more broadly, phenotype risk scores (PheRS) is one method that has been useful to identify undiagnosed diseases (Bastarache et al., 2018, 2019).

To deploy CDS for PRS or ML-driven predictions of treatment response, this chapter summarizes strategies to prepare data to inform CDS; validate and evaluate CDS; explain and interpret CDS; and implement and disseminate CDS. As such applications mature to the point of deployment, there will also need to be sufficient evidence for clinical utility. Drawing from approaches in comparative effectiveness research, pragmatic clinical trials will be useful to understand the relevance of an intervention in routine clinical practice. In addition, when patient subgroups are known, retrospective study designs will be suitable. When assessing clinical utility, it is essential to also understand limitations to identifying and providing guidance to important subgroups, as well as to consider patient preferences.

## Conclusion

While progress has been made toward applying our understanding of pharmacogenes in precision medicine, additional efforts are needed. Strategies in AI and ML are promising to bring together such factors to discover patient subtypes and to predict treatment response for those subtypes. This chapter summarizes several recommendations toward achieving the vision of ML/AI-driven clinical decision support to deliver personalized guidance on drug therapies.

## References

Athreya, A. P., Neavin, D., Carrillo-Roa, T., Skime, M., Biernacka, J., Frye, M. A., Rush, A. J., Wang, L., Binder, E. B., Iyer, R. K., Weinshilboum, R. M., & Bobo, W. V. (2019). Pharmacogenomics-driven prediction of antidepressant treatment outcomes: A machine-learning approach with multi-trial replication. *Clinical Pharmacology and Therapeutics, 106*(4), 855–865. https://doi.org/10.1002/cpt.1482

Bastarache, L., Hughey, J. J., Goldstein, J. A., Bastraache, J. A., Das, S., Zaki, N. C., Zeng, C., Tang, L. A., Roden, D. M., & Denny, J. C. (2019). Improving the phenotype risk score as a scalable approach to identifying patients with Mendelian disease. *Journal of the American Medical Informatics Association, 26*(12), 1437–1447. https://doi.org/10.1093/jamia/ocz179

Bastarache, L., Hughey, J. J., Hebbring, S., Marlo, J., Zhao, W., Ho, W. T., Van Driest, S. L., McGregor, T. L., Mosley, J. D., Wells, Q. S., Temple, M., Ramirez, A. H., Carroll, R., Osterman, T., Edwards, T., Ruderfer, D., Velez Edwards, D. R., Hamid, R., Cogan, J., … Denny, J. C. (2018). Phenotype risk scores identify patients with unrecognized mendelian disease patterns. *Science, 359*(6381), 1233–1239. https://doi.org/10.1126/science.aal4043

Bates, D. W., Auerbach, A., Schulam, P., Wright, A., & Saria, S. (2020). Reporting and implementing interventions involving machine learning and artificial intelligence. *Annals of Internal Medicine, 172*(11), S137–S144. https://doi.org/10.7326/M19-0872

Boland, M. R., Hripcsak, G., Shen, Y., Chung, W. K., & Weng, C. (2013). Defining a comprehensive vero type using electronic health records for personalized medicine. *Journal of the American Medical Informatics Association, 20*(2), e232–e238. https://doi.org/10.1136/amiajnl-2013-001932

Casey, J. A., Schwartz, B. S., Stewart, W. F., & Adler, N. E. (2016). Using electronic health records for population health research: A review of methods and applications. *Annual Review of Public Health, 37*, 61–81. https://doi.org/10.1146/annurev-publhealth-032315-021353

Cho, S. Y., Hong, E. J., Nam, J. M., Han, B., Chu, C., & Park, O. (2012). Opening of the National Biobank of Korea as the infrastructure of future biomedical science in Korea. *Osong Public Health and Research Perspectives, 3*(3), 177–184. https://doi.org/10.1016/j.phrp.2012.07.004

Demner-Fushman, D., Chapman, W. W., & McDonald, C. J. (2009). What can natural language processing do for clinical decision support? *Journal of Biomedical Informatics, 42*(5), 760–772. https://doi.org/10.1016/j.jbi.2009.08.007

Gage, B. F., Eby, C., Johnson, J. A., Deych, E., Rieder, M. J., Ridker, P. M., Milligan, P. E., Grice, G., Lenzini, P., Rettie, A. E., Aquilante, C. L., Grosso, L., Marsh, S., Langaee, T., Farnett, L. E., Voora, D., Veenstra, D. L., Glynn, R. J., Barrett, A., & McLeod, H. L. (2008). Use of pharmacogenetic and clinical factors to predict the therapeutic dose of warfarin. *Clinical Pharmacology and Therapeutics, 84*(3), 326–331. https://doi.org/10.1038/clpt.2008.10

Gibson, G., & Barsh, G. S. (2019). On the utilization of polygenic risk scores for therapeutic targeting. *PLoS Genetics, 15*(4), e1008060. https://doi.org/10.1371/journal.pgen.1008060

Goroll, A. H. (2017). Emerging from EHR purgatory-moving from process to outcomes. *New England Journal of Medicine, 376*(21), 2004–2006. https://doi.org/10.1056/NEJMp1700601

Gottesman, O., Kuivaniemi, H., Tromp, G., Faucett, W. A., Li, R., Manolio, T. A., Sanderson, S. C., Kannry, J., Zinberg, R., Basford, M. A., Brilliant, M., Carey, D. J., Chisholm, R. L., Chute, C. G., Connolly, J. J., Crosslin, D., Denny, J. C., Gallego, C. J., Haines, J. L., … Williams, M. S. (2013). The electronic medical records and genomics (eMERGE) network: Past, present, and future. *Genetics in Medicine: Official Journal of the American College of Medical Genetics, 15*(10), 761–771. https://doi.org/10.1038/gim.2013.72

Guan, Y., Zhang, H., Quang, D., Wang, Z., Parker, S. C. J., Pappas, D. A., Kremer, J. M., & Zhu, F. (2019). Machine learning to predict anti–tumor necrosis factor drug responses of rheumatoid arthritis patients by integrating clinical and genetic markers. *Arthritis and Rheumatology, 71*(12), 1987–1996. https://doi.org/10.1002/art.41056

Guo, T., Xiang, Y. T., Xiao, L., Hu, C. Q., Chiu, H. F. K., Ungvari, G. S., Correll, C. U., Lai, K. Y. C., Feng, L., Geng, Y., Feng, Y., & Wang, G. (2015). Measurement-based care versus standard care for major depression: A randomized controlled trial with blind raters. *American Journal of Psychiatry, 172*(10), 1004–1013. https://doi.org/10.1176/appi.ajp.2015.14050652

Health Level Seven International. (2021). HL7 Fast Healthcare Interoperability Resources (FHIR) standard. https://www.hl7.org/fhir/. Accessed 4 Aug 2021.

Hood, L., & Flores, M. (2012). A personal view on systems medicine and the emergence of proactive P4 medicine: Predictive, preventive, personalized and participatory. *New Biotechnology, 29*(6), 613–624. https://doi.org/10.1016/j.nbt.2012.03.004

Hood, L., & Friend, S. H. (2011). Predictive, personalized, preventive, participatory (P4) cancer medicine. *Nature Reviews Clinical Oncology, 8*(3), 184–187. https://doi.org/10.1038/nrclinonc.2010.227

Kawamoto, K., & McDonald, C. J. (2020). Designing, conducting, and reporting clinical decision support studies: Recommendations and call to action. *Annals of Internal Medicine, 172*(11), S101–S109. https://doi.org/10.7326/M19-0875

Ke, W.-S., Hwang, Y., & Lin, E. (2010). Pharmacogenomics of drug efficacy in the interferon treatment of chronic hepatitis C using classification algorithms. *Advances and Applications in Bioinformatics and Chemistry, 3*, 39–44.

Kim, D., Kim, J. H., & Moore, J. H. (2021). Translational bioinformatics: Integrating electronic health record and omics data. *Pacific Symposium on Biocomputing, 26*, 356–359.

Kohane, I. S. (2011). Using electronic health records to drive discovery in disease genomics. *Nature Reviews Genetics, 12*(6), 417–428. https://doi.org/10.1038/nrg2999

Li, R., Kim, D., & Ritchie, M. D. (2017). Methods to analyze big data in pharmacogenomics research. *Pharmacogenomics, 18*(8). https://doi.org/10.2217/pgs-2016-0152

Linder, J. E., Bastarache, L., Hughey, J. J., & Peterson, J. F. (2021). The role of electronic health records in advancing genomic medicine. *Annual Review of Genomics and Human Genetics, 22*, 219–238. https://doi.org/10.1146/annurev-genom-121120-125204

Lin, E., & Lane, H. Y. (2017). Machine learning and systems genomics approaches for multi-omics data. *Biomarker Research, 5*(1). https://doi.org/10.1186/s40364-017-0082-y

Lundberg, S., Allen, P., & Lee, S.-I. (n.d.). A unified approach to interpreting model predictions. https://doi.org/10.5555/3295222.

Mittelstadt, B., Russell, C., & Wachter, S. (2019). Explaining explanations in AI. In *FAT* 2019 - proceedings of the 2019 conference on fairness, accountability, and transparency* (pp. 279–288). Association for Computing Machinery, Inc. https://doi.org/10.1145/3287560.3287574

Mothilal, R. K., Sharma, A., & Tan, C. (2020). Explaining machine learning classifiers through diverse counterfactual explanations. In *FAT* 2020 - proceedings of the 2020 conference on fairness, accountability, and transparency* (pp. 607–617). Association for Computing Machinery, Inc. https://doi.org/10.1145/3351095.3372850

Office of the National Coordinator for Health Information Technology. (2019). 21st Century Cures Act: Interoperability, Information Blocking, and the ONC Health IT Certification Program. Federal Register. 84 FR 7424. https://www.federalregister.gov/documents/2019/03/04/2019-02224/21st-century-cures-act-interoperability-information-blocking-and-the-onc-health-it-certification. Accessed 4 Aug 2021.

Overby, C., Weng, C., & Haerian, K. (2013). Evaluation considerations for EHR-based phenotyping algorithms: A case study for drug-induced liver injury. *AMIA Joint Summits on Translational Science proceedings, 2013*.

Paquette, A. G., Hood, L., Price, N. D., & Sadovsky, Y. (2020). Deep phenotyping during pregnancy for predictive and preventive medicine. *Science Translational Medicine, 12*(527). https://doi.org/10.1126/scitranslmed.aay1059

Rajkomar, A., Dean, J., & Kohane, I. (2019). Machine learning in medicine. *New England Journal of Medicine, 380*(14), 1347–1358. https://doi.org/10.1056/NEJMra1814259

Ribeiro, M. T., Singh, S., & Guestrin, C. (2018). Anchors: High-precision model-agnostic explanations. In *32nd AAAI conference on artificial intelligence, AAAI 2018* (pp. 1527–1535). AAAI Press. https://aaai.org/Library/AAAI/aaai18contents.php.

Roth, J. A., Bradley, K., Thummel, K. E., Veenstra, D. L., & Boudreau, D. (2015). Alcohol misuse, genetics, and major bleeding among warfarin therapy patients in a community setting. *Pharmacoepidemiology and Drug Safety, 24*(6), 619–627. https://doi.org/10.1002/pds.3769

Sankar, P., & Parker, L. S. (2013). The precision medicine initiative's all of us research program: An agenda for research on its ethical, legal, and social issues. *Journal of the American Medical Informatics Association, 19*, 144–151. https://doi.org/10.1038/gim.2016.183\\n11 \\tWeiskopf NG

Schatz, M. C., Philippakis, A. A., Afgan, E., Banks, E., Carey, V. J., Carroll, R. J., Culotti, A., Ellrott, K., Goecks, J., Grossman, R. L., & Hall, I. M. (2022). Inverting the model of genomics data sharing with the NHGRI Genomic Data Science Analysis, Visualization, and Informatics Lab-space. *Cell Genomics, 2*(1), 100085.

Sieberts, S. K., Zhu, F., García-García, J., Stahl, E., Pratap, A., Pandey, G., Pappas, D., Aguilar, D., Anton, B., Bonet, J., Eksi, R., Fornés, O., Guney, E., Li, H., Marín, M. A., Panwar, B., Planas-Iglesias, J., Poglayen, D., Cui, J., … Mangravite, L. M. (2016). Crowdsourced assessment of common genetic contribution to predicting anti-TNF treatment response in rheumatoid arthritis. *Nature Communications, 7*. https://doi.org/10.1038/ncomms12460

Tcheng, J., Bakken, & Bates, D. W. (2017). *Optimizing strategies for clinical decision support - National Academy of Medicine*. National Academies Press.

Topol, E. J. (2019). High-performance medicine: The convergence of human and artificial intelligence. *Nature Medicine, 25*(1), 44–56. https://doi.org/10.1038/s41591-018-0300-7

Torkamani, A., Wineinger, N. E., & Topol, E. J. (2018). The personal and clinical utility of polygenic risk scores. *Nature Reviews Genetics, 19*(9), 581–590. https://doi.org/10.1038/s41576-018-0018-x

Uher, R. (2016). Person-centered measurement-based care for depression. *World Psychiatry, 15*(3), 238–239. https://doi.org/10.1002/wps.20363

Vayena, E., Blasimme, A., & Cohen, I. G. (2018). Machine learning in medicine: Addressing ethical challenges. *PLOS Medicine, 15*(11), e1002689. https://doi.org/10.1371/journal.pmed.1002689

Vimal, M., Devi, W. P., & McGonigle, I. (2021). GenomeAsia100K: Singapore builds national science with Asian DNA. *East Asian Science, Technology and Society: An International Journal, 15*(2), 238–259. https://doi.org/10.1080/18752160.2021.1896138

Wachter, S., Mittelstadt, B., & Russell, C. (2017). Counterfactual explanations without opening the black box: Automated decisions and the GDPR. *SSRN Electronic Journal*. https://doi.org/10.2139/ssrn.3063289

Weiskopf, N. G., & Weng, C. (2013). Methods and dimensions of electronic health record data quality assessment: Enabling reuse for clinical research. *Journal of the American Medical Informatics Association, 20*(1), 144–151. https://doi.org/10.1136/amiajnl-2011-000681

Yurkovich, J. T., Tian, Q., Price, N. D., & Hood, L. (2020). A systems approach to clinical oncology uses deep phenotyping to deliver personalized care. *Nature Reviews Clinical Oncology, 17*(3), 183–194. https://doi.org/10.1038/s41571-019-0273-6

CHAPTER SEVEN

# Usability and human-centered design

**Harry Hochheiser**[1], **Katrina M. Romagnoli**[2]
[1]Department of Biomedical Informatics, University of Pittsburgh, Pittsburgh, PA, United States
[2]Department of Translational Data Science and Informatics, Geisinger, Danville, PA, United States

## Contents



*Clinical Decision Support for Pharmacogenomic Precision Medicine*
ISBN: 978-0-12-824453-1
https://doi.org/10.1016/B978-0-12-824453-1.00009-9

## Introduction

Clinical pharmacogenomics is, in many ways, an exercise in applied informatics. Of course, the underlying understanding of how pharmacokinetics and pharmacodynamics are impacted by various genetic variants is the foundation that makes pharmacogenomics possible, but the translation of these insights into patient care rests on the availability of accurate information presented clearly to help clinicians and patients make informed and appropriate decisions. To meet these needs, information systems must account for the breadth and complexity of information about genetic variants, pathways, medications, comorbidities, drug interactions, and evidence regarding outcomes of potential interventions, ideally by bridging differences in perspectives between nurses, physicians, pharmacists, and their patients. This is not easy to do well: abundant experience with electronic health records and other clinical information systems provides detailed insight into the difficulties associated with building useable clinical tools, including the myriad ways that well-intentioned efforts can go wrong (Boonstra et al., 2021; Howe et al., 2018; Mazur et al., 2019; Middleton et al., 2013; Moy et al., 2020; Ratwani et al., 2018).

Avoiding these pitfalls requires sustained and detailed efforts aimed at understanding the background and perspectives of the different classes of individuals involved in pharmacogenomic decision-making, using insights from this process to guide the design of effective tools (often with the participation of impacted individuals), and evaluating and refining designs based on experiences with using the resulting designs in both laboratory and field settings. These efforts draw on more than 40 years of experience in usability, human-centered design, and human—computer interaction, often complemented by even older results from early work in ergonomics and human factors (Jacobs et al., 2017; Schuler & Namioka, 1993). These approaches provide no guarantees of success, but human-centered design techniques can help guide and contextualize efforts, potentially providing useful insights when systems don't meet goals, thus enabling a (perhaps frustratingly gradual) evolution toward effective system design.

As the name implies, "human-centered design" (often referred to as HCD) addresses these questions by putting the user first. Our exploration of HCD will start with a discussion of some of the factors that make pharmacogenomics a challenging problem for user-centered design. We will introduce the types of people involved in pharmacogenomic decision-making, and how their unique perspectives make different—and often contradictory—demands on system design. Having identified these groups, we will consider specific tasks and goals and related information gaps that might challenge users and lead to breakdowns along the way. We will then outline various user-centered design techniques and research methods appropriate for collecting information about users and their goals, analyzing the results of these explorations to build models explaining user goals, translating insights from those models into designs, evaluating the effectiveness of those designs, and iterating through multiple attempts. Appropriate research methods and artifacts (tools and visualizations suitable for communicating important findings) will be described. Examples from recent efforts in pharmacogenomics will illustrate how these techniques have been applied, and what might be learned from those earlier experiences.

## Why is pharmacogenomics usability hard?

The usability of pharmacogenomics is hard, in no small part, because usability in clinical information systems is hard. Despite a history of medical informatics dating back to the mid-1970s and years of experience with Electronic Medical Record (EMR) systems in medical care environments ranging from small medical practices to nationwide systems, reports of usability difficulties, excessive burdens on clinicians, and potentially dangerous workarounds are commonplace (Boonstra et al., 2021; Howe et al., 2018; Mazur et al., 2019; Ratwani et al., 2018). Often attributed to the complexities of clinical workflows, navigation challenges, and mismatches between clinician goals and systems requirements (Moy et al., 2020; Roman et al., 2017; Smith & Koppel, 2014), these difficulties have led to calls for increased attention to usability issues (Middleton et al., 2013).

Although there may be many causes for these difficulties, the complexities of medical care are arguably among the most intractable. The nuances of complex patient cases requiring the interpretation and integration of multiple data sources are not easily captured in a few simple pull-down menus. Furthermore, the vast amount of complex data that might be needed to reach appropriate clinical conclusions and interpretations presents additional challenges in the design of tools which present information effectively, without overwhelming or confusing clinicians.

These difficulties are widespread and well-documented for relatively "routine" aspects of medical care. For pharmacogenomics, these problems arise in the complexities of understanding the implications of multiple genomic variants, regulatory pathways, related medications, and evidence for effectiveness of proposed interventions. Furthermore, the relative novelty of pharmacogenomics in practice suggests informatics tools must go beyond helping clinicians arrive at appropriate decisions, to actively helping convey the increasingly nuanced and sophisticated fluency with the underlying concepts necessary for effective patient-focused prescribing and management of medications in the context of relevant genomic alterations. Ideally, these tools will build shared understanding between physicians, nurses, pharmacists, patients, and caregivers, ultimately leading to greater efficiency and precision in medication.

## Information

When making a clinical decision related to pharmacogenomics, the clinician must gather, interpret, and consider disparate information items including but not limited to the medication or medications in question, the variant or variants which affect those medications, the implication of the combination of the patient's specific variant and medications, other side effects and interactions of the rest of the medications prescribed, and evidence regarding the best course of action. These decisions are complicated: a variant might necessitate a lower dose of one medication but avoiding other medications altogether. For example, *CYP2C19* poor metabolizers are exposed to higher quantities of (es)citalopram compared to extensive and ultra-rapid metabolizers, and typically see clinical effects from the medication at lower doses. However, *CYP2C19* poor metabolizers do not activate the pro-drug clopidogrel well, and thus do not receive clinical benefit from it. Alternative medications might be appropriate. Similarly, different genes can affect the same medication in different, opposing ways. Other variants might require selection of alternative medications due to the risk of dangerous hypersensitivity reactions. The information is accessible, but requires support from trustworthy, easy to access information sources at the point of care. Keeping track of the relationship between genes, variants, phenotypes, and treatment decisions for medications requires a mental burden better handled by clinical decision support tools.

Pharmacogenomics can also interfere with clinical workflows and processes. Unlike other tests, pharmacogenomic genotyping does not yet have a rapid turnaround time. Clinicians are often hesitant to order pharmacogenomic tests until multiple treatments have failed, but preemptive pharmacogenomic testing is not typically covered by insurance plans. The patient needs to be treated, and the risks of delaying treatment likely exceed the risks of being given a suboptimal medication or dose for the patient's specific genotype.

Even when all information is available in a timely manner, moving from test results and medication information to interpretation and planning is not straightforward. Pharmacogenomics information may be presented in complex EMRs or medication administration systems, alongside dozens of other data points. Details of the implications of variants and potential impacts might not be immediately available, forcing clinicians to review local policies, external resources, or even primary literature. Although computerized decision support systems might help, mixed experiences with these tools suggest that simply telling clinicians what to do is not sufficient. Effective tools must help clinicians build the understanding needed to have confidence in their ability to use informed understanding of underlying complexities to guide decision-making. By focusing on the clinician perspective, human-centered design approaches can shed light on relevant gaps and suggest how those gaps might be met.

### *Introduction to human-centered design*

Human-centered design offers principles and methods to incorporate the end user into the entire design process, with the goal of the end product being a highly useable, highly functional tool that supports the needs of the user by fitting seamlessly into their workflow. Human-centered design consists of developing understanding of and empathy with the people for whom you are designing; generating ideas; building one or more prototypes to evaluate; evaluating the usability and effectiveness of those prototypes with the users; and iteratively improving to the point of a final product.

A number of models have been developed to articulate this process. One example of a model of human-centered design, the IDEO model of innovation (Brown, 2019), places innovation at the intersection of desirability (what the user wants), feasibility (what is possible to build), and viability (what is required for success). Therefore, understanding each of these aspects is critical for design success of a tool that is intended to solve a problem for a human. Bruce Hanington's model of design research is a simple three-step process moving from exploration to generation and evaluation. Each of these activities involves its own set of methods, which are often used in an iterative and overlapping manner. The first phase is *Explore*, where the researcher/designer conducts ethnographic, qualitative research to understand the user and the context in depth. The second phase is *Generate*, where the researcher/designer prototypes possible solutions that might address what is learned in the *Explore* phase. The final phase is *Evaluate*, where the researcher/designer conducts research to understand whether or not the prototypes succeed at solving the problems identified in the *Explore* phase, how useable they are, and how users feel about them. The process may repeat iteratively, generating prototypes of greater fidelity and maturity until the solution may be implemented in the real world (Hanington, 2007).

Following these processes requires developing a deep understanding of the needs and perspectives of stakeholders. Drawing on sociological and ethnographic research traditions, designers use a variety of techniques to understand not only the immediate challenge being addressed in any design effort but also the underlying goals, contexts, and constraints. Information needs, specific subtasks, workflows, barriers, social interactions, and personal preferences might constrain feasible solutions. As models of these complex situations are inherently incomplete, iterative evaluation and refinement of the designer's understanding, and of proposed tool designs, is almost always necessary.

In the next section, we will provide a brief introduction to each of these aspects of human-centered design as well as commonly used methods and artifacts to obtain this understanding, with discussion of how these efforts have been applied to pharmacogenomics systems.

## Stakeholders

Any computer system will directly or indirectly impact multiple different groups of individuals, each having their own goals and needs. Understanding these varied perspectives is a critical first step toward building designs that consider the needs of all of those who are touched by a design. Known as "stakeholders" these groups are usually discussed in terms of *roles* individuals or groups will play in the context of specific environments. This distinction reflects the fluidity of each individual's perspective as they shift between different tasks and roles. A trained physician might take on multiple roles as they interact with a pharmacogenomics system, including (but not limited to) *clinician,* as they interact with patients; *manager,* as they review records for patients in a group; *researcher* as they use clinical data to study impacts of pharmacogenomics on patient outcomes; *patient* as they discuss their own medication with their physicians; and possibly *caretaker* as they review medications for young children or elderly parents. Each of these roles—even when embodied in the same individual—will have unique goals, tasks, and information needs.

Through inclusion of indirect impacts, the concept of stakeholder is broader than that of *user*. One needn't be sitting at the keyboard of touching the screen to be a stakeholder in a system. This is a particularly important concern for pharmacogenomics systems, as patients definitely have an interest in decisions, even if they never interact directly with electronic medical records, medication administration systems, drug reference databases, or other relevant computing tools. Similarly, researchers, healthcare administrators, pharmacy and therapeutics committees, and health plans might also be considered to be stakeholders. Of course, not every system need involve every potential stakeholder: a thorough stakeholder analysis can be a helpful tool for explicitly identifying which stakeholders are out of scope, thereby constraining (and perhaps simplifying) designs.

## Goals

Understanding stakeholders' macro and micro goals is critical. Key questions include (but are not limited to):

What are the stakeholder's broad goals? What problems are they trying to solve? What are the specific goals related to the project at hand? How is success measured?

Answers to these questions will vary but capturing as many of these as possible will contribute to an understanding of what users and stakeholders are trying to accomplish and how specific design solutions might best support those goals. For example, in pharmacogenomics clinical decision support, an organization might be interested in decreasing the amount of time, on average, for patients receiving a new anticoagulant therapy to reach the in-range International Normalized Ratio (INR). Pharmacogenomic-informed clinical decision support may help clinicians reach that goal, with the right information, designed in the right way. Inquiries with relevant staff might suggest that individual clinicians want to increase the number of patients who undergo pharmacogenomic panel testing, but they struggle to convince patients of the value of having the blood test. Shaping the problem space and understanding it clearly will support the goal of designing a solution that helps the users and stakeholders achieve their goals.

## Information needs, tasks, and workflow

To accomplish these goals, users will have specific information needs, tasks, and workflows. Information needs can be specific, such as outlining a recommended course of action for a patient with a specific pharmacogenomic genotype and phenotype. Information needs can also exist at an organizational level, such as estimating how many patients in the practice have a specific pharmacogenomic genotype and phenotype, and how many of them are currently on a drug affected by the genotype. A clinician might want information about individual genotypes and specific medications to make a decision about a patient on their schedule that day, while a health system administrator might want the cohort data relating variants to medications, outcomes, and costs to make a decision about population healthcare for all of their patients.

Tasks are the means that must be taken to achieve desired ends—stakeholder goals. Some tasks might be necessary—pharmacogenomics might not be able to proceed without review of a genetic test, while others might exist to meet procedural needs. Established processes—particularly bureaucratic processes—may include multiple requirements for documentation or approval which do not contribute to stakeholders' end goals. Reframing these steps in terms of the underlying end goal may lead to the identification of opportunities for redesigning and streamlining workflow. In an example familiar to many participants in the US healthcare system, obtaining written prior approval for a specialist visit might be seen as an intrinsic goal, implemented as a means of achieving both the patient's goal of receiving appropriate care and the insurer's goal of limiting unnecessary expenses. The realization that the written approval is not a necessary task might lead to the implementation of electronic sign-off on referrals, thus providing necessary accountability without troublesome paperwork.

Tasks can also be hierarchical, consisting of multiple subtasks. *Hierarchical Task Analysis* is the process of decomposing complex tasks into successively more refined subtasks (Stanton, 2006). In the case of computer systems, this can often mean down to the level of individual keystrokes and button clicks. Although this approach is useful for understanding current work practices, researchers should be careful to note that these analyses will describe work as it is, not as it should be. Task analysis is often used as a starting point for the work redesign.

Tasks and information needs are often combined—explicitly or implicitly—to form complex *workflows* characterized by dependencies on external information and actions and partial orders between tasks. In clinical settings, best practices and system protocols often dictate what a clinician must do in the course of their work. Workflows in EHRs can be prescriptive and complicated; time is short. Tasks and information needs can be intertwined. Understanding what tasks users must complete and the workflow necessary to complete those tasks is critical in designing a solution that both meets their needs and does not interfere or compete with their necessary tasks and workflow.

## Context

In addition to specific roles and goals, the use of a pharmacogenomic system is often constrained by the time, location, and specific motivations for any given interactions. For example, an intensive care clinician reviewing pharmacogenomic results during morning rounds might experience time pressure and competing demands for their attention which might not be present if they were completing the same tasks in their office later in the working day. These constraints define the contexts in which information needs are addressed, tasks are completed, and workflows guide action toward achieving user goals. Contexts contain multiple levels of factors and influences that influence work, from broader cultural factors to others specific to a given place and time. The context of a large academic research hospital system in a major city in the United States is very different from the context of a small, independent practice in a rural setting in an Asian country. Context can include the culture of the community and the organization; the tools available to the users; the patterns of clinical practice; even the time schedules of the organization. For instance, if two organizations comparable in every way except their use of different EHR systems, their context is different in an important way. Understanding these differences can be critical for successful system design.

## Human-centered design research methods for pharmacogenomic CDS

In the next section, we introduce a few methods for conducting Human-Centered Design, in the context of pharmacogenomics clinician decision support implementation. These methods are not exhaustive—other methods exist that may be better suited for a specific implementation project. Additional resources are cited at the end of this chapter for further reading.

## Method: stakeholder mapping

Identification of stakeholders is often the first step in gathering task, workflow, and contextual information needed to build a successful system. A commonly used method for capturing stakeholder information is *stakeholder mapping:* the construction of visual maps articulating the key roles of participants who might have a stake in the outcome. Stakeholder mapping sessions gather team members to brainstorm and identify types of users for a proposed tool (i.e., pharmacists, clinicians, nurses, etc.) and other stakeholders who have a vested interest in the outcome (i.e., health system administrators, EHR developers, etc.). Other approaches include capturing insights from contextual interviews, designed to elicit from individuals a description of work processes and how they might relate to a proposed system. These processes can be used to identify various roles participants might play, with roles mapping onto stakeholder groups (Beyer & Holtzblatt, 2016). Alternatively, scenario-based design suggests the creation of written stories (scenarios) describing how different users might accomplish key goals. These scenarios can then be interrogated to extract perspectives of classes of stakeholders (Rosson & Carroll, 2002).

Whichever approach is used, HCD experts will generally strive to capture as many of these roles, groups, or specific people as possible, and to model how they are related to each other. These definitions can often provide a starting point for identifying partners in the development of the various phases of the design project (ideating, prototyping, evaluating, etc.) As with all other aspects of user-centered design, stakeholder identification is often an iterative process of ongoing refinement and validation.

Understanding the relationships between relevant stakeholder groups is an important step in determining which groups should be involved in design processes, and how. These can be mapped to a Power Plot (Fig. 7.1). Stakeholders will have varying levels of *interest* in the design outcome, as well as varying levels of *power,* meaning the ability to influence the outcome. For instance, the primary end user of a solution might have the greatest interest in the design outcome, because it impacts their workflow, but a lower level of power over the outcome. A hospital administrator may have a higher level of power over the outcome, but less interest in specific details. A clinical informatics administrator might have both a high level of interest in and a high level of power over the outcome. One potentially useful approach to understanding these dynamics involves plotting individuals on a graph linking the level of interest (X-axis) and to the level of power available to that

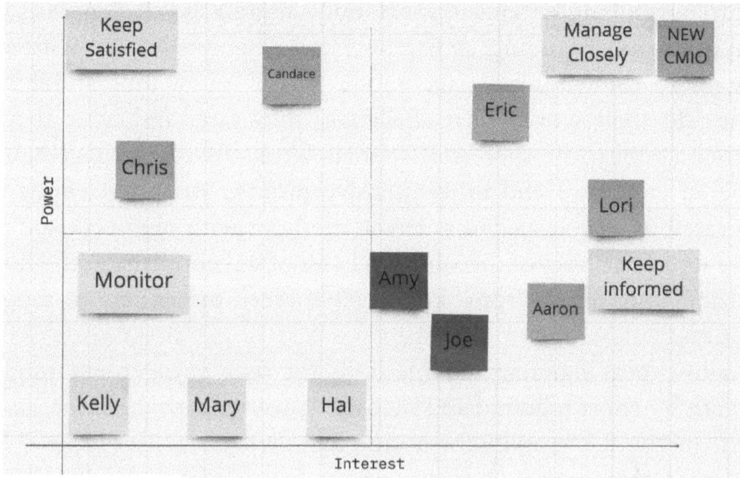


**Figure 7.1** Stakeholder interest-power plot. This figure shows a representative stakeholder interest-power plot.

individual (Y-axis). Each individual's position in the plot provides insight into which people might need close management (those with high interest and high power), some degree of monitoring (low interest and low power), versus those who might simply need to be kept informed (high interest, low power), or otherwise satisfied (low interest, high power). Alternative approaches involve the construction of cultural diagrams (Beyer & Holtzblatt, 2016), graphical representations describing relationships, and information flows between stakeholders, including potential areas where breakdowns might occur.

Going beyond identification of specific individuals or roles, stakeholder mapping often includes detailed descriptions of information needs, goals, workflows, and contextual factors. These descriptive details might be listed in a stakeholder description, or they might be described in narrative form, through *persona* descriptions painting a detailed picture of the stakeholder and their challenges (Rosson & Carroll, 2002). Although many of these details might not be well-understood at the outset, these maps can be valuable tools for identifying and prioritizing individuals for participation in user research inquiries that will help fill the gaps.

## User research

Preliminary efforts at identifying stakeholders and understanding tasks and goals provide a starting point for developing the in-depth understanding needed to developing systems to successfully address the challenges at hand. To build this understanding, researchers and designers engage with stakeholders, often using multiple techniques aimed at eliciting insights into how they do their work, what challenges they face, and what they will need from a tool to be successful at their work. As user engagement can often be an expensive and time-intensive process, successful efforts often plan to carefully match specific research methods with the needs and availability of representatives of the various classes of stakeholders, in the hope of maximizing the input from key stakeholders presenting a range of perspectives.

Roughly speaking, research methods for user engagement might be sorted both by effort required and by overall utility, with the understanding that these methods exemplify the notion that there is no "free lunch." Relatively low-cost methods such as surveys can be useful, but the closed nature of surveys limits their utility, as answers are only given to questions that are asked. Moving on to more labor-intensive approaches such as interviews and ethnographic methods provides more detailed insight, at significantly increased expense in both time and resources.

## Method: surveys

Surveys are relatively low-cost research tools for collecting data from a large number of people at one time. They can be distributed electronically using a variety of tools (such as Qualtrics, SurveyMonkey, REDcap, Google Forms, among many others) or by paper or telephone. Electronic and phone surveys allow for faster development, distribution, and analysis compared to paper surveys, but may exclude some participants. Using standard, validated survey instruments such as Likert scale responses (Likert, 1932), Net Promoter Scale (Reichheld, 2003), System Usability Scale (Brooke, 1986), SUPR-Q (Sauro, 2015), and others is a powerful way to collect information. However, if specific information is required that does not map directly to a validated instrument, developing a survey de novo may be necessary. Another consideration is how many "free text" responses are necessary or desirable: although yes/no, multiple-choice and Likert-scale questions can be easily tabulated and analyzed through descriptive and inferential approaches, analyzing free text survey data is time consuming and requires rigorous qualitative data analysis methods. Surveys can also be limiting, as respondents generally cannot answer questions which are not asked. Other challenges in survey design include developing appropriate questions, identifying an appropriate sample of the population in questions and acquiring an appropriate number of responses—ideally more than 50% of those asked.

### *Examples from literature*

When successful, the payoff can be substantial, as surveys can provide a broad understanding of the attitudes and perceptions of representative subgroups of key stakeholder populations. Several studies have used surveys to understand various aspects of clinician opinions on pharmacogenomics, including early studies of clinician knowledge and use of pharmacogenomic testing (Stanek et al., 2012), to more recent inquiries into perceptions of barriers and benefits to the application of genomic medicine (Krebs & Milani, 2019; Owusu-Obeng et al., 2014).

## Method: stakeholder interviews and focus groups

Surveys can be useful for providing broad feedback, but researchers and designers often find more depth from a smaller number of participants is needed to develop the rich understanding needed for system development. To gather these insights, HCD practitioners take the next step of asking stakeholders to talk about various aspects of their work, through interviews and focus groups.

Although seemingly straightforward, interviews and focus groups present challenges in design, execution, and analysis. Stakeholder maps might provide the first step, in the form of a preliminary list of individuals, or at least stakeholder groups who might be invited to participate. The next step will likely be to plan the content of the session. While structured interviews guided by a static list of questions are possible, most interviews are *semistructured,* using a structured set of questions to be asked formally as starting points for open-ended discussions as the interviewer digs deeper into comments made by the interviewee. Interviewees can suggest additional people to be interviewed, potentially including representatives of otherwise unidentified stakeholder groups.

Focus groups are essentially group interviews. In gathering multiple informants, potentially representing different stakeholder groups, in a single session, focus groups provide the possibility of capturing rich details from multiple perspectives in a relatively short period of time. Groups are often a particularly attractive option when recruiting executives, clinicians, and other professionals who are difficult to schedule. However, focus groups can be challenging, as dominant personalities or influential people (often the same person) might drown out less assertive individuals (the "sunflower effect"), or power dynamics might limit participants' willingness to speak candidly. Focus groups are often best conducted by a team of researchers, allowing one person to lead the group while another takes notes and perhaps a third watches group dynamics and works to ensure that all voices are heard. Sessions can be audio and/or video-recorded for re-play and transcription. One on one interviews are most effective to receive honest, unbiased, candid views from a variety of people with different perspectives.

### *Examples from literature*

A recent example of the use of semistructured qualitative interviews is a study of clinicians to inform the implementation of pharmacogenomics into general practice in the United Kingdom by Rafi et al. They interviewed 18 clinicians and identified barriers, challenges, and opportunities related to the implementation of pharmacogenomics. Most notably was the clinicians' concern about the cost-effectiveness of pharmacogenomics in primary care, which might be alleviated by the rise in direct-to-consumer genetic testing (Rafi et al., 2020).

## Method: contextual inquiry and think-aloud studies

Interviews can be useful, but they are limited by the time and place where they are conducted. Discussions occurring across a meeting-room table (or a web conference) removed from how and where the work is completed run the risk of relying on the informant's memory and a simplified version of the tasks at hand. A more informative and also more time-intensive approach is to learn about the user's work as they do it. *Contextual inquiries* are a type of interview combined with observations, where the participant is both interviewed and observed in their relevant context, such as a clinic or pharmacy, as they conduct relevant tasks. In these sessions, the interviewer will ask the participant to show them how the work is done, as if the participant was teaching the interviewer to do the job (Beyer & Holtzblatt, 2016). Another option is to conduct *think-aloud evaluation,* which entails asking the participant to complete a task while speaking out loud about what they are doing, why they are doing it, and what they are thinking about and feeling as they do it (Lazar et al., 2017).

### *Examples from literature*

The authors of this chapter conducted contextual inquiries with pharmacists to inform the design and development of a pharmacogenomics information resource for pharmacists. That work identified the information needs pharmacists experience when making medication prescribing recommendations in general and specific to pharmacogenomics, which informed the data model for the information resource. Another key finding was the importance of having an easy to understand, ideally graphical representation of the relationship between pharmacogenomic genotypes, phenotypes, medications, and recommendation, as well as a clear understanding of the strength of evidence behind that recommendation (Boyce et al., 2013; Romagnoli, 2015; Romagnoli et al., 2016, 2017).

## Method: design ethnography

Design ethnography is a process of immersing oneself in the context and environment of the user. Though it comes from the fields of anthropology and sociology, design ethnography for the purposes of understanding user needs and context is distinct from true ethnography. Spending time with and among the users for whom a solution is designed is critical to understanding the details of their work. For instance, scheduling time in clinics and pharmacies with providers and pharmacists will help the researcher understand the tools being used; the daily flow and experience of the clinic or pharmacy; the challenges encountered; other stakeholders or staff the user encounters, among many other details. During ethnography, the researcher can collect data of various kinds including, but not limited to, photos of workstations and the environment (de-identifying PHI as appropriate), notes of observations, quotes from the staff and patients, and workflow diagrams (Lazar et al., 2017).

## Mixed methods

Observations and interviews gather rich, contextual data that answers such as "What are the goals and tasks of my user group?", "What challenges are they experiencing that I might be able to improve?", "Why are they doing that?". Qualitative research is rich but time-consuming, data point heavy but participant light. Pairing small "N" qualitative methods with large "N" quantitative methods, like surveys, to conduct *mixed methods research* is a powerful way to collect compelling data to inform design decisions.

Some researchers begin with a survey, then use the results of the survey to inform the development of in-depth qualitative research methods, to understand the *why* behind the survey responses. Other researchers begin with qualitative research to explore the domain, identify themes and hypotheses for further research, then validate their findings with a larger population in the form of a survey. Both are valid choices. Findings from a smaller number of pharmacists about their preferences in how pharmacogenomics information is presented in their workflow can be turned into a survey that is sent to a larger number of pharmacists, asking for their feedback about a few suggested options for pharmacogenomics information displays. The results of the survey might validate the findings from the qualitative work, or they might lead to iterative refinement and improved understanding. This combination can be a powerful tool for informing design decisions.

### *Examples from literature*

Devine et al. conducted a mixed methods usability study, capturing both qualitative feedback from cardiologists and oncologists as they used their prototype CPOE prescribing tool with pharmacogenomics alerts, while also measuring their actions on screen and responses to surveys. They identified opportunities for iteratively improving the pharmacogenomics prescribing tool. Blagec et al. conducted qualitative semistructured interviews to evaluate the usability and acceptability of a mobile pharmacogenomics prescribing tool, followed by surveys with pharmacogenomics experts and physicians and pharmacists to understand the best way to present pharmacogenomics information and attitudes toward the system. They found that the system was accepted by users and identified additional ways to provide information support about pharmacogenomics (Blagec et al., 2016).

## Analysis

Data collected in human-centered design research efforts is usually described as being either quantitative or qualitative. Typically resulting from surveys, *quantitative data* involve counts and numeric measures—the number of participants agreeing or disagreeing with a statement, the average (and standard deviation) of answers to a question on a 5-point subjective scale, etc. Interviews, focus groups, observations, and ethnographies generate *qualitative data*—textual descriptions provided by stakeholders or captured by observers. Thorough human-centered research efforts value each of these data types for the insights they provide—quantitative data provides insight into the magnitude of a reaction or effect, and qualitative data provides insight into underlying causes and mechanisms. Whichever data type is involved, appropriate analysis techniques are necessary to generate robust and defensible results.

Quantitative data can be analyzed through descriptive and inferential (parametric and nonparametric) statistical techniques. Descriptive statistics are often useful for providing insight into the population of stakeholders participating in a study. Particularly for systems which may impact diverse groups of individuals—such as patients potentially impacted by pharmacogenomics—human-centered studies will want to identify individuals of different education level, ages, genders, ethnicities, or even understanding of key medical concepts. Descriptive statistics, generally provided in tables or charts, can illustrate the range of variation, and to identify any concerns about populations that might be implicitly or explicitly biased in ways that might influence interpretation of the data.

Inferential statistics can be used to compare groups of participants. These comparisons are particularly useful to understand differences between stakeholders—for example, a well-designed survey might ask both patients and clinicians to express their perceived level of understanding of key pharmacogenomics concepts. Statistical comparisons of answers to questions can be used to explore the extent and magnitude of any observed differences between the groups. For summative studies of functional systems, inferential statistics are often used to compare task completion times and error rates between proposed systems designs (see Evaluation below).

Inferential statistics can be dangerous if used inappropriately, particularly for surveys. Although surveys based on responses from a carefully sampled, representative subset of a population provide answers that can be generalized to the population as a whole, effective sampling is often a challenge. If your sample is more opportunistic (known as a "convenience sample"), you will want to clearly state these limitations and to avoid overly broad statements regarding how your results apply to the broader population.

Qualitative data analysis involves a very different, but no less rigorous, set of techniques. When analyzing feedback from interviews, notes from observations, or insights from ethnographic inquiries, the initial question is not "how much?" or "how many?" but instead "what is going on?" Rather than developing charts and graphs, qualitative analyses aim to describe the work, goals, contexts, and stakeholders through the identification of recurring themes and stories.

The first step in a qualitative analysis involves reading through notes and transcripts to identify and categorize comments or observations of interest. Known as *coding,* this process involves some combination of two key forms (Creswell & Poth, 2017; Saldaña, 2021). *Emergent coding*, otherwise known as "letting the data speak for itself," or *open coding,* starts from a blank slate, with codes added as needed to describe observations of interest. As themes emerge, these codes might be grouped into categories, building a hierarchy of concepts and concerns. As analyses continue, identified codes and themes might be used to construct graphical diagrams (Beyer & Holtzblatt, 2016) and/or scenario descriptions (key aspects of the situation) (Rosson & Carroll, 2002). Alternatively, a priori coding starts from a theory or model thought to be relevant to the problem at hand. Concepts from the theory of interest are used as the major categories in the coding process.

Whether emergent or a priori, qualitative coding can be accompanied by accepted practices designed to guarantee appropriate rigor. As the concepts included in a coding effort become reasonably clear, researchers will generally develop a *codebook* defining relevant codes and providing examples of comments associated with each concept. This codebook can then be used by multiple coders who will code a given subset of the data. High agreement in ratings, measured through interrater reliability metrics, will provide confidence in the reliability of the coding practices (Lazar et al., 2017). Widely used qualitative analysis software tools provide support for automating these processes. Additional rigor can be provided through triangulation—the process of using input from multiple informants, possibly collected through different means, to support any conclusions that might be drawn from the data. Qualitative analysis can also involve some quantitative aspects. For example, counts of the number or percentage of participants raising a concern might provide some insight into the importance of that issue.

Reporting the results of qualitative studies can be a challenge. One common approach is to summarize main categories, providing example quotes as needed to illustrate important themes. This approach was used in a study of barriers to pharmacogenomics implementation in general practice in the United Kingdom. Based on analysis of 18 semistructured interviews with clinicians, researchers identified seven themes (possible utility, patient factors, pharmacogenomics education, social implications, cost-effectiveness, and informatics), providing examples of each (Rafi et al., 2020).

## Design

Although useful, quantitative data and qualitative summaries may not provide the depictions of work necessary to guide design. Researchers and practitioners have developed strategies for using visual artifacts to ground designers, decision makers, developers, and other stakeholders on the context and experience specific to the organization in which tools might be implemented. Discussed below are three examples human-centered design artifacts frequently used to guide design.

## Artifact: experience maps and journey maps

An *experience map* captures the current state of an end-to-end experience (such as using a service) from the perspective of the user or customer. Experience maps are typically generic to an experience not tied to a specific product or service (such as the experience of being a cancer patient), while journey maps are typically specific to the experience of using a particular product or service (such as using an app designed to support cancer patients). Both capture information related to what the typical person is doing, thinking, and feeling at different stages in an experience, what tools they may be using, as well as other relevant information for that particular experience, tool, or user. A visual component often graphs the emotional or affective experience. The goals of the experience map, according to Adaptive Path's *Guide To Experience Mapping* (Adaptive Path, 2015), are to

- Create a shared frame of reference around the user experience;
- Build organization knowledge of behaviors and needs across channels;
- Identify specific areas of opportunity to drive ideation and innovation;
- Distribute key user insights an in easy-to-understand form;
- Further organizational evolution toward user-centered thinking.

The data used to generate experience maps and journey maps include observations, contextual inquiries, interviews, and surveys. They are often organized using *swim lanes*: horizontal rows that contain information about the experience over time.

### *Example from literature*

Hall et al. developed a Cancer Experience Map to elevate the voice of the cancer patient to help content developers understand the complex cancer patient experience and design cancer support videos and other mobile health technology for cancer patients (Hall et al., 2015).

## Artifact: service blueprint

Service blueprints resemble experience maps, but they articulate the current state from both the perspective of the user or customer and the internal organization. They are like workflow diagrams in that they visualize processes to accomplish a task. They have swim lanes like experience maps. A key component of a service blueprint the depiction of processes happening "in front of house" (as in, the front of house at a restaurant where customers are served), processes happening "in back of house" (as in, the back of house at a restaurant where the food is prepared), and the interface of information and stakeholders between the two (as in, the waitstaff communicating between the customer and the kitchen). A service blueprint is an excellent artifact when it is important to articulate the flow of data and resources combined with the user workflow. Service blueprints help identify opportunities for innovation and improvement by illustrating complex or poorly integrated processes and flows of information. A service blueprint of a pharmacogenomics CDS tool could illustrate the sources of information the CDS uses, the flow of that information, and the tasks the user completes using the CDS tool or tools. Service blueprints are context specific: they are not generic to a common experience but precisely visualize the flow and infrastructure of a specific service or product implementation. In illustrating flows of information between different stakeholders and sequences of actions, service blueprints are similar to flow and sequence diagrams used in contextual design practices (Beyer & Holtzblatt, 2016).

### *Example from literature*

Silva et al. conducted a qualitative study in two phases to create and implement a service blueprint illustrating comprehensive medication management in a rheumatology ambulatory clinic. The project identified multiple points at which the service could break down, resulting in a suboptimal patient experience. They also identified needs to support services to improve service delivery and patient experience (Silva et al., 2021). Litchfield et al. conducted a mixed methods study (surveys and interviews) to understand the return of test result communication process in England, then developed a service blueprint to illustrate the process and identify opportunities for improvement (Fig. 7.2).

## Artifacts: personas and user scenarios

Personas are aggregate representations of a typical user or customer, or a specific segment or type of user or customer. They are fictional but based in data. They evoke empathy for the user and a shared understanding across stakeholders in *who* the tool or service is serving, what their *challenges* are, and what they are trying to *accomplish* as a user. They are often developed and used in conjunction with other artifacts, such as service blueprints or experience maps.

Personas can have a visual component, such as a stock photo, a drawing, or an icon representing the persona or segment. The primary content of the persona is demographic information, as well as details and context relevant to the design project. For instance, a design team implementing

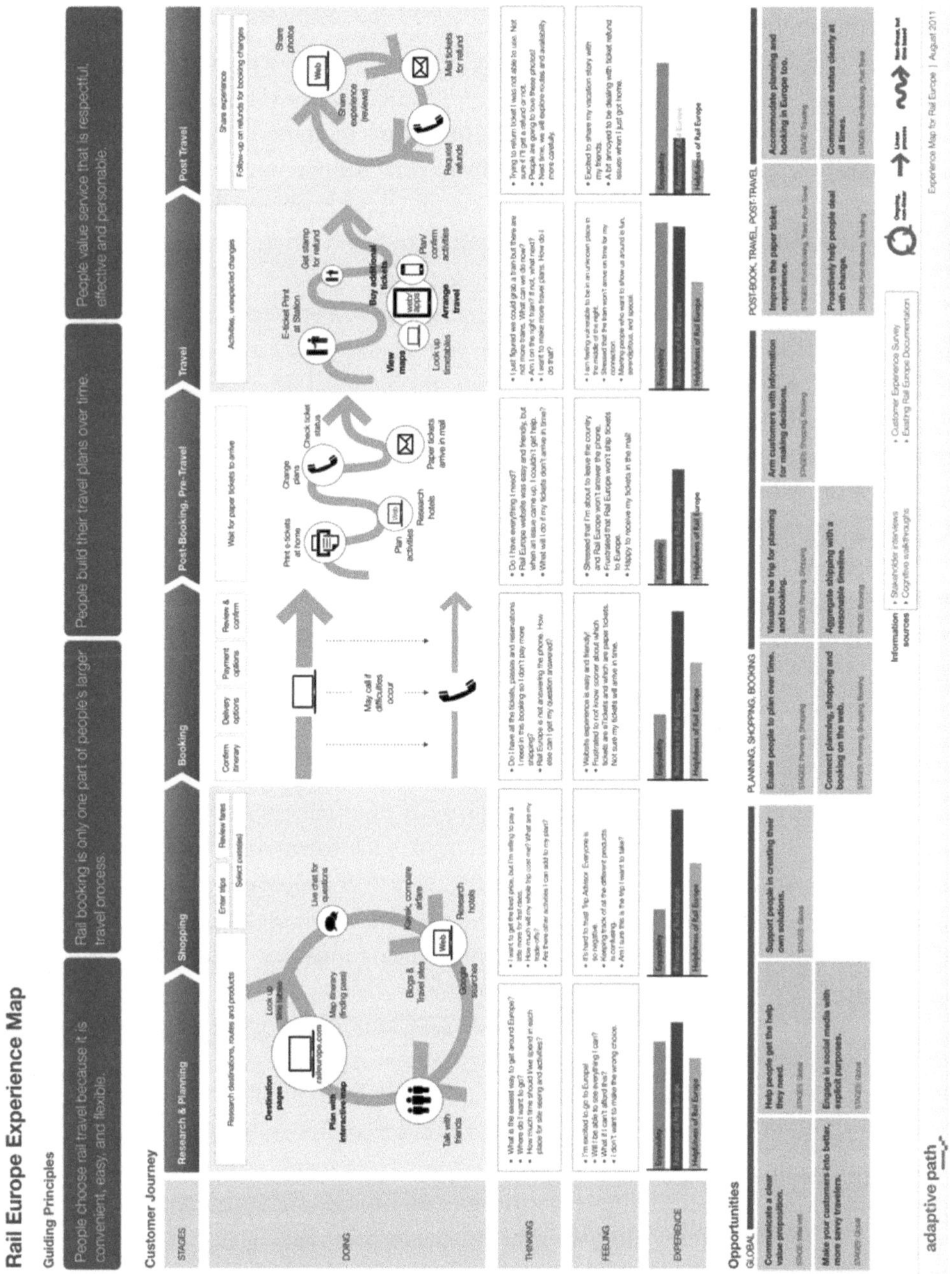


Figure 7.2 Service blueprint of test results communication in England. Service blueprint.

pharmacogenomics CDS might have a persona for a hospital-based pharmacist, one for a retail-based pharmacist, and one for a physician, representing three user groups who are expected to be frequent users of the tool. It is important to avoid slipping into overgeneralizations or stereotypes of demographic groups—as that might be inadvertently offensive or too simplistic of a representation of the users—while remaining true to the data and simple enough to be easily understood by someone not familiar with the work. The data used to generate personas can come from both qualitative and quantitative sources, such as interviews, contextual inquiries, observations, surveys, and demographic data.

User scenarios (or use cases) are stories, often linked to personas, depicting how a fictional user interacts with a service or tool. Like personas, they are fictional but grounded in data. They illustrate the ideal situation in which a person engages with a service or tool, what their expectations, needs, and goals are, with the intention of creating a shared understanding across stakeholders of what the goals and expected functionality of the service or tool is or will be in the future.

### *Example of personas and user scenarios in pharmacogenomics*

In dissertation research by Romagnoli, the results of qualitative inquiries were used to develop a use case persona of a composite pharmacist and use case scenarios. The persona was written as a composite of the characteristics discussed and observed during the qualitative inquiries. The user scenarios (Figs. 7.3–7.5) describe the likely use cases, information needs, and workflows of pharmacists who will engage with pharmacogenomics information. The persona and use cases were used to inform the design and development of a prototype pharmacogenomics information resource to meet the pharmacists' needs and support their information seeking behavior when managing medications with pharmacogenomics implications (Romagnoli, 2015).

### *Building solutions*

Having collected and analyzed both qualitative and quantitative data, and having used the results of those processes to build experience maps, journey maps, service blueprints, scenarios, and other representations of the problem at hand, human-centered design practitioners will move on to the next key challenge—designing new tools to improve upon prior practices (if any) to solve key problems. This process involves iteration through several steps providing increasingly greater levels of detail and fidelity, culminating in a fully functional tool.

| Description of a Composite Pharmacist Based on Interviews | |
|---|---|
| **Demographics** | Margaret, Clinical Pharmacist, female, 40 years old , 15 years of experience at large academic health system |
| **Work style** | As Margaret's work is fast-paced, she is often multi-tasking. To meet these demands, Margaret places a high value on using easily accessible, trustworthy information to make fast decisions. |
| **Job Tasks** | Margaret's job requires her to round on patients, answer questions from physicians, and make recommendations about medication treatment plans. She interacts with clinicians of all types, including physicians, nurses, and other pharmacists. |
| **Experience with information resources** | Margaret has worked with many different information systems, including electronic medical records, tertiary drug information resources, and literature searching tools. She uses all of them regularly. Her favorite tertiary resources are Lexicomp™ and Dynamed™, but she also uses Micromedex™ and Up-To-Date™. She checks everything against the FDA product label. |
| **Preferences in information resources** | Trustworthy information from a familiar resource is important to Margaret. She is particularly inclined to trust familiar resources that answer her questions quickly and succinctly, with the ability to get more details if she needs them. She wants to know the source of the information, and prefers primary sources and FDA information. |
| **Values and goals at work** | Margaret takes her job very seriously. As she is particularly interested in evidence-based medicine, she strives to stay apprised of the latest developments in the literature. Her colleagues consider her very well informed, and trust her judgment and advice. Margaret values having and using cutting edge medical knowledge; educating herself and her colleagues on new developments in both medical science and clinical practice. She has a strong interest in continuous improvement of medical care and information resources. |
| **Experience with Pharmacogenomics** | She is aware of pharmacogenomics and the effects genotypes can have on different drugs. She would like to incorporate pharmacogenomics information and data into his practice, because she wants to both optimize patient care and safety, and protect her license. Genotyping is not currently being performed in her health system but she expects it to become more common in the future, and thinks it will improve clinical care and outcomes. |

**Figure 7.3** Use case 1. General pharmacogenomics question during hospital rounds.

| | |
|---|---|
| **Tasks** | Margaret is on rounds, using her mobile device to find and share medication information with the clinicians. She also has access to the EMR system. |
| **Information Needs** | One of the residents wants to switch a patient's medication to clopidogrel, and asks Margaret what the group should know about it. She wants information about the drug, specifically the presence of important genotypes that could predict the patient's response before she fills the prescription, whether the patient has been tested for any of those genotypes, and what the results are, if the test has been performed, what the results of the test might mean, and if/how the treatment should be changed based on the results. |
| **Information Seeking** | She has tried using other resources to answer these questions, such as the hospital's intranet resources, Micromedex™, Lexicomp™, and Up-to-Date™, but they don't answer her questions. She has heard about other resources, such as PharmGKB and GeneTests, but when she looked at them previously, she did not understand the information presented to her enough to use it. |

**Figure 7.4** Use case 2. Clinical decision based on patient genotype.

| | |
|---|---|
| **Tasks** | Margaret is verifying medication orders. She sees in the EMR that the patient has a genotype that can result in an undesirable response in patients taking abacavir. |
| **Information Needs** | She wants information about how severe the response is, how likely it is that the patient will experience an adverse medication event as a result, what the adverse medication event is likely to be, whether she should recommend the physician switch to a different treatment, and what the change ought to be (different drug, or different dose). |
| **Information Seeking** | The information Margaret is looking for is not available in the EMR. She has to leave this system to find any information about pharmacogenomics, and she has a hard time understanding the information she finds in other resources. She finds a recommendation in the product label, which she considers difficult to use, though she trusts the information. |

**Figure 7.5** Use case 3. Clinical decision based on patient genotype.

Earlier stages in the process will generate broad-stroke depictions of possible workflows and processes, focused on the goal of developing a consensus around key elements of the proposed system. Examples include flowcharts and diagrams indicating steps in a process; activity scenarios describing how a tool will be used, by whom, and when (Rosson & Carroll, 2002); or cartoon-style storyboards telling stories about key interactions. These artifacts communicate how the proposed tool might be used. As consensus is reached, these early models will be fleshed out to include specific descriptions of user interface design elements and specific, lower-level depictions of key tasks. They are iteratively enhanced and refined, leading to low-fidelity paper or online prototypes depicting how interactions might appear on a screen.

Contextual factors including which users will be interacting with a tool and where they will be using it must be considered at this phase in design. Pharmacogenomics tools designed to be used in the clinic by physicians working at a desktop, or perhaps a computer-on-wheels, might involve information, terminology, and interactions not appropriate for a patient-facing, web-based tool meant to be accessed on mobile devices. Carefully consideration of these user and contextual factors will help ensure all needs are met, or to determine when multiple distinct designs are necessary.

Throughout these design efforts, designers will be in frequent discussion with stakeholders and potential users. These exchanges are vitally important for ensuring proposed designs meet their intended goals and are likely to be well-accepted by those most impacted. Stakeholder comments can be invaluable, as "no, that's not quite right" or similar comments can suggest times when iterative revision or complete reimagination of proposed designs is needed. The earlier in the design and development process this feedback can be collected, the easier and less resource-intensive it will be to incorporate it. As designs progress toward paper prototypes, participants might propose alternative screen layouts or interactions. One common technique involves the use of sticky notes to build paper prototypes, as notes might be rearranged, removed, or added to enhance the system design.

## Evaluation

As earlier discussion of design has hinted, evaluation can—and should—begin early in the process and continue iteratively throughout. Early reviews of proposed workflows should involve only a small set of expert informants, while paper and eventually functional prototypes provide opportunities for increasingly broader sets of both designers and users to participate in evaluation activities.

Early evaluations provide detailed feedback evaluating the fitness of the proposed interface for the relevant tasks and goals, including suggested changes aimed at addressing critical difficulties. Known as *formative* evaluations, these efforts can begin when paper or low-fidelity prototypes are available and continue iteratively through the development of functional styles. Formative evaluations generally fall into one of two camps: expert reviews and usability studies.

Expert reviews ask user interaction experts and domain experts to critically examine user interfaces to identify potential problems in design and/or workflow. Although main variants exist, these reviews come in two main flavors. *Heuristic evaluations* ask experts to evaluate how well an interface adheres to published guidelines for effective interface design. Experts will examine displays and interactions, noting violations that fail to uphold these best practices, producing a list of problems with subjective judgments of severity and possibly proposed revisions. These evaluations can be accomplished inexpensively and have been found to find a significant portion of major usability difficulties (Nielsen, 1993).

*Cognitive walkthroughs* involve the construction of user descriptions and the simulated execution of key tasks by those users, focusing on repeated inquiry as to whether the simulated user would be likely to successfully complete the task at hand. These walkthroughs provide insights into potential workflow breakdowns that might require redesign (Mahatody et al., 2010).

Usability studies ask users to attempt to complete appropriate tasks using the prototype or end product (Blagec et al., 2016; Devine et al., 2014; Romagnoli et al., 2017). Early usability studies might focus on identifying design gaps through "think-aloud" exercises. These studies ask participants to verbalize their goals, interpretations of the interfaces, and actions, as they complete a task. Inconsistencies between user statements and aspects of the proposed designs might a need for design revisions to account for potential misunderstandings (Lazar et al., 2017). As design elements converge and confidence in the proposed solutions increases, usability studies might be enhanced with preliminary quantitative measures of task completion time and correctness, providing some concrete validation of success.

Toward design completion, focus will shift toward evaluations demonstrating the effectiveness of the tool for completing key goals, often in comparison to existing tools or alternative designs. These *summative* evaluations generally collect quantitative measures of key factors of success such as task completion time, error rate, and subjective satisfaction. Less commonly studied, but often important factors include learnability and memorability (Shneiderman & Plaisant, 2010). Timing and correctness measures can often be captured through instrumented versions of the proposed software (Lazar et al., 2017) with subjective satisfaction measured through widely used tools such as the System Usability Scale (Brooke, 1986).

Analyses of results from summative evaluations will be determined by the goals of the study. Evaluations involving a single-proposed design are often used to determine acceptability relative to minimal performance standards. Although these studies can involve statistical analyses, descriptions of performance relative to benchmarks are often sufficient. Comparisons between systems, or between design alternatives, often go somewhat further, using inferential statistics in the hope of demonstrating statistically significant differences between tools.

Most of these studies, including early summative studies, will likely be conducted in simulated laboratory environments using carefully constructed tasks. If these efforts prove fruitful, designers can turn their attention toward studies of the tool in action as intended. Field studies (Friedman & Wyatt, 2006) must be undertaken carefully, as intervening in care practice with actual patients can have negative impacts. Many studies will focus on time-limited investigations in specific care environments before expanding scope to a broader audience. These studies also require careful design of quasi-experimental or randomly controlled protocols, as gaps in experimental designs can lead to potential confounds that might impact the generalizability of results. Successful field studies involve extensive planning and the cooperation of clinicians and researchers to execute effectively.

## Human-centered design in pharmacogenomics: selected examples

Human-centered design has been a staple in pharmacogenomics research for several years. A 2014 study from the University of Washington combined think-aloud studies with measurements of activity on a prototype system to identify possible categories of improvements to a proposed pharmacogenomics decision support design (Devine et al., 2014). A subsequent study by this group used simulated tasks in a prototype pharmacogenomics clinical decision support system to show that the use of the system increased the proportion of clinicians seeing advantages of the pharmacogenomics decision support, with many physicians making significant changes to prescribing patterns (Overby et al., 2015). A 2016 study by Blajec et al. evaluated a mobile pharmacogenomics CDS tool for physicians and found it was well-received by clinicians, and identified opportunities to improve the communication of pharmacogenomics information (Blagec et al., 2016). A 2017 study by the authors of this chapter evaluated a pharmacogenomics information tool for pharmacists designed using human-centered design (Romagnoli et al., 2017). A 2017 review identified more than 30 publications detailing descriptions of user-interactions with pharmacogenomics decisions support tools. Other efforts have focused on specific challenges, such as individual medications (Nguyen et al., 2019).

## Conclusion

As a complex, dynamic, and highly impactful field at the intersection of multiple domains, pharmacogenomics presents a significant challenge to clinicians, patients, and other healthcare administrators. Human-centered design provides techniques for understanding stakeholder needs and goals and applying that understanding to the development of effective tools for pharmacogenomics education and clinical decision support. Lessons learned from previous efforts, including the need to provide clear depictions of the implications of pharmacogenomics information, are likely to remain relevant as increasing experience with these approaches identifies some successes and unearths further difficulties.

It is important to note that although useful, human-centered design is no panacea. Numerous decision support tools have taken all the right steps and yet still failed. Learning from failures may be as important, if not more important than demonstrating success. Effective research efforts will explore new ideas, learning from bold failures rather than settling for minor successes. In that spirit, human-centered design can also help, as ambitious efforts might dare to engage clinicians and patients together in efforts that use pharmacogenomics to reimagine the complexities of how clinicians and patients interact to discuss medications.

## References

Adaptive Path. (2015). *Adaptive Path's guide to experience mapping*. Blurb, Incorporated. https://books.google.com/books/about/Adaptive_Path_s_Guide_to_Experience_Mapp.html?id=u4EDswEACAAJ&source=kp_book_description.

Beyer, H., & Holtzblatt, K. (2016). *Contextual design: Design for life (interactive technologies)* (p. 530). https://www.sciencedirect.com/book/9780128008942/contextual-design.

Blagec, K., Romagnoli, K. M., Boyce, R. D., & Samwald, M. (2016). Examining perceptions of the usefulness and usability of a mobile-based system for pharmacogenomics clinical decision support: A mixed methods study. *PeerJ, 4*, e1671. https://doi.org/10.7717/peerj.1671

Boonstra, A., Jonker, T. L., van Offenbeek, M. A. G., & Vos, J. F. J. (2021). Persisting workarounds in electronic health record system use: Types, risks and benefits. *BMC Medical Informatics and Decision Making, 21*(1). https://doi.org/10.1186/S12911-021-01548-0

Boyce, R., Freimuth, R., Romagnoli, K., T, P., H, H., & PE, E. (2013). Toward semantic modeling of pharmacogenomic knowledge for clinical and translational decision support. *AMIA Joint Summits on Translational Science Proceedings*, 28—32. https://pubmed.ncbi.nlm.nih.gov/24303292/.

Brooke, J. (1986). SUS - a quick and dirty usability scale. In *Usability evaluation in industry* (pp. 189—194).

Brown, T. (2019). *Change by design, revised and updated* (p. 294).

Creswell, J. W., & Poth, C. N. (2017). *Qualitative inquiry & research design: Choosing among five approaches* (4th ed., p. 459). SAGE Publications, Inc..

Devine, E. B., Lee, C. J., Overby, C. L., Abernethy, N., McCune, J., Smith, J. W., & Tarczy-Hornoch, P. (2014). Usability evaluation of pharmacogenomics clinical decision support aids and clinical knowledge resources in a computerized provider order entry system: A mixed methods approach. *International Journal of Medical Informatics, 83*(7), 473—483. https://doi.org/10.1016/j.ijmedinf.2014.04.008

Friedman, C. P., & Wyatt, J. (2006). *Evaluation methods in medical informatics* (2nd ed., p. 386). Springer.

Hall, L. K., Kunz, B. F., Davis, E. V., Dawson, R. I., & Powers, R. S. (2015). The cancer experience map: An approach to including the patient voice in supportive care solutions. *Journal of Medical Internet Research, 17*(5), e132. https://doi.org/10.2196/jmir.3652

Hanington, B. M. (2007). Generative research in design education. *International Association of Societies*, 1—15. https://www.mendeley.com/catalogue/89e7e959-8d53-371f-a4e1-a75dc7625dae/?utm_source=desktop&utm_medium=1.19.4&utm_campaign=open_catalog&userDocumentId=%7Ba1ebd460-f329-4e8c-90ff-ad7d5cfee1ac%7D.

Howe, J. L., Adams, K. T., Hettinger, A. Z., & Ratwani, R. M. (2018). Electronic health record usability issues and potential contribution to patient harm. *Journal of the American Medical Association, 319*(12), 1276–1278. https://doi.org/10.1001/JAMA.2018.1171

Jacobs, S., Elmqvist, N., Cohen, M., Plaisant, C., & Shneiderman, B. (2017). *Designing the user interface: Strategies for effective human-computer interaction.* Pearson Education.

Krebs, K., & Milani, L. (2019). Translating pharmacogenomics into clinical decisions: Do not let the perfect be the enemy of the good. *Human Genomics, 13*(1), 39. https://doi.org/10.1186/S40246-019-0229-Z

Lazar, J., Feng, J. H., & Hochheiser, H. (2017). *Research methods in human-computer interaction.*

Likert, R. (1932). A technique for measurement of attitudes. *Archives of Psychology, 140,* 1–55.

Mahatody, T., Sagar, M., & Kolski, C. (2010). State of the art on the cognitive walkthrough method, its variants and evolutions. *International Journal of Human–Computer Interaction, 26*(8). https://doi.org/10.1080/10447311003781409

Mazur, L. M., Mosaly, P. R., Moore, C., & Marks, L. (2019). Association of the usability of electronic health records with cognitive workload and performance levels among physicians. *JAMA Network Open, 2*(4), e191709. https://doi.org/10.1001/JAMANETWORKOPEN.2019.1709

Middleton, B., Bloomrosen, M., Dente, M. A., Hashmat, B., Koppel, R., Overhage, J. M., Payne, T. H., Rosenbloom, S. T., Weaver, C., & Zhang, J. (2013). Enhancing patient safety and quality of care by improving the usability of electronic health record systems: Recommendations from AMIA. *Journal of the American Medical Informatics Association, 20*(e1), e2–e8. https://doi.org/10.1136/AMIAJNL-2012-001458

Moy, A. J., Schwartz, J. M., Elias, J., Imran, S., Lucas, E., Cato, K. D., & Rossetti, S. C. (2020). Time-motion examination of electronic health record utilization and clinician workflows indicate frequent task switching and documentation burden. *AMIA Annual Symposium Proceedings, 2020,* 886. https://www.ncbi.nlm.nih.gov/pmc/articles/PMC8075533/.

Nguyen, K. A., Patel, H., Haggstrom, D. A., Zillich, A. J., Imperiale, T. F., & Russ, A. L. (2019). Utilizing a user-centered approach to develop and assess pharmacogenomic clinical decision support for thiopurine methyltransferase. *BMC Medical Informatics and Decision Making, 19*(1). https://doi.org/10.1186/s12911-019-0919-4

Nielsen, J. (1993). *Usability engineering* (p. 358). Morgan Kaufmann.

Overby, C. L., Devine, E. B., Abernethy, N., McCune, J. S., & Tarczy-Hornoch, P. (2015). Making pharmacogenomic-based prescribing alerts more effective: A scenario-based pilot study with physicians. *Journal of Biomedical Informatics, 55,* 249–259. https://doi.org/10.1016/j.jbi.2015.04.011

Owusu-Obeng, A., Weitzel, K. W., Hatton, R. C., Staley, B. J., Ashton, J., Cooper-Dehoff, R. M., & Johnson, J. A. (2014). Emerging roles for pharmacists in clinical implementation of pharmacogenomics. *Pharmacotherapy, 34*(10), 1102–1112. https://doi.org/10.1002/phar.1481

Rafi, I., Crinson, I., Dawes, M., Rafi, D., Pirmohamed, M., & Walter, F. M. (2020). The implementation of pharmacogenomics into UK general practice: A qualitative study exploring barriers, challenges and opportunities. *Journal of Community Genetics, 11*(3), 269–277. https://doi.org/10.1007/s12687-020-00468-2

Ratwani, R., Savage, E., Will, A., Arnold, R., Khairat, S., Miller, K., Fairbanks, R., Hodgkins, M., & Hettinger, A. (2018). A usability and safety analysis of electronic health records: A multi-center study. *Journal of the American Medical Informatics Association, 25*(9), 1197–1201. https://doi.org/10.1093/JAMIA/OCY088

Reichheld, F. F. (2003). The one number you need to grow. *Harvard Business Review, 81*(12), 46–54. https://pubmed.ncbi.nlm.nih.gov/14712543/.

Romagnoli, K. (2015). *User-centered design and evaluation of a pharmacogenomics information portal to support clinical decision-making.*

Romagnoli, K. M., Boyce, R. D., Empey, P. E., Adams, S., & Hochheiser, H. (2016). Bringing clinical pharmacogenomics information to pharmacists: A qualitative study of information needs and resource requirements. *International Journal of Medical Informatics, 86*, 54–61. https://doi.org/10.1016/j.ijmedinf.2015.11.015

Romagnoli, K., Boyce, R. D., Empey, P. E., Ning, Y., Adams, S., & Hochheiser, H. (2017). Design and evaluation of a pharmacogenomics information resource for pharmacists. *Journal of the American Medical Informatics Association, 24*(4), 822–831. https://doi.org/10.1093/jamia/ocx007

Roman, L., Ancker, J., Johnson, S., & Senathirajah, Y. (2017). Navigation in the electronic health record: A review of the safety and usability literature. *Journal of Biomedical Informatics, 67*, 69–79. https://doi.org/10.1016/J.JBI.2017.01.005

Rosson, M. B., & Carroll, J. M. (2002). *Usability engineering: Scenario-based development of human-computer interaction* (p. 422).

Saldaña, J. (2021). *The coding manual for qualitative researchers* (4th ed.). SAGE Publications Ltd.

Sauro, J. (2015). SUPR-Q: A comprehensive measure of the quality of the website user experience. *Journal of Usability Studies, 10.*

Schuler, D., & Namioka, A. (1993). Participatory design: Principles and practices (p. 334). Hillsdale, New Jersey: Thomas Erlbaum Associates.

Shneiderman, B., & Plaisant, C. (2010). *Designing the user interface: Strategies for effective human-computer interaction* (p. 606).

Silva, H., Gonzaga do Nascimento, M., de Morais Neves, C., Oliveira, I., Cipolla, C., Batista de Oliveira, G., Yone de Almeida, N., & Ramalho de Oliveira, D. (2021). Service blueprint of comprehensive medication management: A mapping for outpatient clinics. *Research in Social & Administrative Pharmacy.* https://doi.org/10.1016/J.SAPHARM.2021.01.006

Smith, S. W., & Koppel, R. (2014). Healthcare information technology's relativity problems: A typology of how patients' physical reality, clinicians' mental models, and healthcare information technology differ. *Journal of the American Medical Informatics Association, 21*(1), 117–131. https://doi.org/10.1136/AMIAJNL-2012-001419

Stanek, E. J., Sanders, C. L., Taber, K. A. J., Khalid, M., Patel, A., Verbrugge, R. R., Agatep, B. C., Aubert, R. E., Epstein, R. S., & Frueh, F. W. (2012). Adoption of pharmacogenomic testing by US physicians: Results of a nationwide survey. *Clinical Pharmacology & Therapeutics, 91*(3), 450–458. https://doi.org/10.1038/CLPT.2011.306

Stanton, N. (2006). Hierarchical task analysis: Developments, applications, and extensions. *Applied Ergonomics, 37*(1), 55–79. https://doi.org/10.1016/J.APERGO.2005.06.003

CHAPTER EIGHT

# Clinical considerations for precision medicine clinical decision support

**James M. Stevenson[1], April Schultz[2,3], Amanda Massmann[2,3], Joel Van Heukelom[2,3], Casey Overby Taylor[4,5,6]**

[1]Department of Medicine, Division of Clinical Pharmacology, Johns Hopkins University School of Medicine, Baltimore, MD, United States
[2]Department of Internal Medicine, University of South Dakota Sanford School of Medicine, Sioux Falls, SD, United States
[3]Sanford Imagenetics, Sanford Health, Sioux Falls, SD, United States
[4]Department of Medicine, Division of General Internal Medicine, Johns Hopkins University School of Medicine, Baltimore, MD, United States
[5]Department of Biomedical Engineering, Johns Hopkins University, Baltimore, MD, United States
[6]Institute for Computational Medicine, Johns Hopkins University, Baltimore, MD, United States

## Contents



*Clinical Decision Support for Pharmacogenomic Precision Medicine*
ISBN: 978-0-12-824453-1
https://doi.org/10.1016/B978-0-12-824453-1.00005-1

## Introduction

Information, in all forms, needs to be readily accessible to be clinically meaningful. It has been reported that physicians spend an average of 16 minutes within the electronic health record (EHR) for each encounter they have, with much of that time in chart review, documentation, and ordering (Overhage & McCallie, 2020). A poll by the RAND Corporation found that EHRs had both positive and negative impacts on physician's professional satisfaction, with poor usability, time consuming data entry, and inefficiencies listed among the negatives (Friedberg et al., 2014). When implementing clinical decision support (CDS), one must harness the positive attributes of the EHR while cutting the negatives by making alerts prompt, actionable, and efficient. Doing so can aid in optimizing medication therapy based on integrated genomic information. Keeping the frontline healthcare provider in mind, this chapter will address considerations for the design and maintenance of effective precision medicine CDS.

## CDS design: the five rights

In some ways, drug—gene interactions are similar to drug—drug interactions. Patient-specific factors (a genetic polymorphism or concomitant medication) affect the pharmacokinetics or pharmacodynamics of a medication, and thus the likelihood of beneficial effects and adverse drug reactions. Drug—gene and drug-drug interactions can have varying levels of severity and supporting evidence. Because of these conceptual parallels, drug-gene interaction CDS can often resemble drug—drug interaction CDS. While drug—drug interaction checking is one of the most common applications of CDS in the medication management process, alert fatigue and dissatisfaction with these measures are common (Osheroff et al., 2009). Individuals designing CDS for drug—gene interactions should be mindful of these concerns and optimize CDS for maximum end user satisfaction and treatment outcomes.

In other ways, drug—gene interactions are conceptually different than drug—drug interactions. Drug—gene interactions require a laboratory test to detect the patient-specific factor (genetic polymorphism). Because a patient's genotype is stable throughout the lifetime, pharmacogenomic testing can be performed in a reactive or preemptive model. Reactive testing refers to triggering testing upon ordering a medication that could be affected by pharmacogenomics. In this setting, prescribers will typically choose a treatment based on clinical factors, and then modify the agent or dose if needed after the pharmacogenomic results become available. The other option is preemptive testing, in which pharmacogenomic testing is done prior to medication initiation. While conceptually this is preferable, the a priori identification of patients likely to start a pharmacogenomic medication in the future can be difficult. Further, third-party reimbursement requirements tend to favor reactive testing. In either preemptive or reactive paradigms, pharmacogenomic testing can consist of a single gene of interest, or multiple genes can be assayed simultaneously (Duarte et al., 2021). Typically, this involves the use of a genotyping microarray that provides a "panel" of pharmacogenomic results. Some institutions perform panel-based testing in a reactive model (for the drug—gene interaction of immediate relevance), but provide results for additional pharmacogenomic genes assayed on the panel preemptively (Duarte et al., 2021; Liu et al., 2021).

There is a clear need to install CDS in the context of preemptive pharmacogenomic results. Without CDS, prescribers throughout the health system may not know that relevant pharmacogenomic results exist, or that the patient had been tested. However, CDS is also important in the more limited context of reactive, single gene testing. Pharmacogenomics of drug metabolizing enzymes can be used to guide treatment for dozens of medications prescribed in various medical specialties or settings. For example, a *CYP2C19* genotype initially ordered by a cardiologist for assistance in selecting an antiplatelet agent after percutaneous coronary intervention can be informative when the patient's primary care provider initiates therapy with the antidepressant citalopram many years later. The prescriber for these two treatment decisions is not the same, and without CDS the primary care provider may not be aware of the years-old order and result for *CYP2C19*. One study of patients that underwent *CYP2C19* genotyping in the context of percutaneous coronary intervention found that more than half were prescribed another medication affected by *CYP2C19* genotype in the following year (Beitelshees et al., 2020). This also highlights the issue of portability of pharmacogenomic results across healthcare systems: if the results are only available within one healthcare system, valuable information that could be used to make clinical decisions may not be available if patients receive care at an institution from another healthcare system.

One strategy for designing effective pharmacogenomic CDS is to use the "CDS Five Rights" framework developed by the Agency for Healthcare

Research and Quality (Osheroff et al., 2009). Not to be confused with the "Five Rights of medication use" (right patient, drug, dose, route, and timing), the CDS Five Rights are specific to optimizing CDS interventions. The components of the CDS Five Rights are (1) the right information, (2) the right person, (3) the right CDS intervention format, (4) the right time in the workflow, and (5) the right channel.

## Right information

While there is much excitement surrounding the use of pharmacogenomic information to both improve drug efficacy and reduce the risk of adverse events, many providers do not feel comfortable with utilizing pharmacogenomic results or where to find guidelines (Rahawi et al., 2020; Smith et al., 2020). Due to this unfamiliarity, CDS plays an important role in guiding the prescribing of medications with defined recommendations based on drug–gene variants. Although the majority of medical, pharmacy, nursing, and health science schools have added pharmacogenomics to their didactic curriculum, there still remains a gap in education (Nicholson et al., 2021). Ideally, the CDS should succinctly describe the drug–gene relationship, provide recommendations, and link to references for further information. Drug–gene relationships that have been elevated by the FDA, such as the use of codeine or tramadol in CYP2D6 ultra-rapid metabolizers (UMs), are not yet widespread knowledge among many pediatricians (Rahawi et al., 2020), which further highlights the need for integrated CDS within the EHR. This is an interesting finding, especially in light of the FDA issuing a black box warning in February of 2012.

Generally speaking, the information conveyed by CDS should be evidence-based, relevant, and suitable to guide action (Osheroff et al., 2009). Increasing the specificity and relevance of the CDS can reduce the risk of alert fatigue. Ideally, the content should be constrained with no superfluous information. Additional supporting information may be optionally accessible via hyperlinks or InfoButtons to educational materials or a laboratory report (Freimuth et al., 2017). In the context of providing CDS to clinically minded end users, the interpretation and treatment recommendation are generally more important than the actual diplotype result or the complete list of variants that were tested. While diplotype and variants tested should be available to end users, the assumption that most clinicians will not consider these data for most decisions suggests a hyperlink to the detailed laboratory result may be appropriate.

Defining the possible medication outcome from a drug–gene interaction may be helpful to end users. Especially in the case of a prodrug (a drug that requires bioactivation within the body to its active form) such as codeine, if the CDS only recommends using a different medication because the patient is a CYP2D6 UM, one can see how this could be misinterpreted as having an alternative recommendation due to decreased efficacy (the incorrect notion that UMs should metabolize the medication more rapidly and require higher doses). It may be clearer and more effective to define that UMs have increased conversion to an active metabolite and are at increased risk for serious side effects within the text of the CDS.

Providing the right information for pharmacogenomic-guided decision-making can include nongenetic information as well. Some pharmacogenomic recommendations are specific to treatment indication. CPIC guidelines for antiplatelet selection are primarily intended for patients undergoing percutaneous coronary intervention and may not apply to other clinical scenarios in which clopidogrel is being considered (e.g., peripheral arterial occlusive disease) (Scott et al., 2013). Pharmacogenomic algorithms intended for clinically dosing warfarin include indication and numerous nongenetic factors such as age, weight, smoking status, and drug interactions (Gage et al., 2008; Klein et al., 2009). Providing information on these clinical variables in the CDS text—or less preferably a disclaimer about when the recommendation should and should not be applied—is certainly pertinent information for appropriate clinical decision-making.

Of course, what information needs to be conveyed to end users should not be solely decided by the CDS or precision medicine teams. An iterative process using input and feedback from end users should be used to provide concise yet sufficient information in the CDS application. Even nontechnical aspects of the text will differ based on end user background and preferences—the wording, "increased risk of ototoxicity," may be sufficient for some users, but others may prefer, "2.3-fold increased risk of ototoxicity." Finally, teams designing CDS around pharmacogenomic concepts should be mindful that the "right information" is likely to evolve as guidelines are updates, understanding of genotype–phenotype associations evolve, and nomenclature changes.

## Right person

CDS can be targeted toward prescribers, pharmacists, nurses, caretakers, patients, and other participants in the medication use process (Osheroff et al., 2009). Quite obviously, CDS should be designed with the end user in mind.

In defining the "right person" or end user, generally one wants to identify the individuals who can act upon the information at hand. For pharmacogenomic-informed prescribing, this should encompass providers who will be prescribing or reviewing medication orders. Depending on the healthcare setting and practice model, this could include physicians, physician assistants, nurse practitioners, and/or pharmacists. Providing CDS about pharmacogenomic-guided treatment may be a lower priority for nonprescribing nurses, given the prescribing provider and pharmacist should have already been informed of any drug–gene interactions earlier in the medication management process.

Another consideration is whether preserving autonomy for end users is pinnacle, or if a referral to a specialist or consultation is preferred. While this consideration should go into selecting the "right information," it also matters for defining end users. One institution described an informatics tool that detected unreadable pharmacogenomic test results (e.g., a PDF from a third-party laboratory) and triggered a manual review for result validation and manual documentation of discrete results (Caraballo et al., 2017). This type of task requires advanced knowledge of pharmacogenomics and does not fit into typical clinic visit, medication ordering, or medication verification workflows. Thus, the receipt of this type of alert should be limited to specified end users with the requisite training and time commitment.

CDS for nonclinical end users (e.g., patient-facing or patient-centered CDS) is possible as well. These types of individuals may be the appropriate audience for tools intending to share pharmacogenomic results with providers outside of their usual healthcare system (Blagec et al., 2018; Blagec et al., 2016). Another possibility is that patients and caregivers are end users for CDS that advises more careful monitoring for adverse drug reactions in the outpatient setting. One concern about the sharing of pharmacogenomic results with patients is that patients may misinterpret results and modify or stop treatment without clinician oversight (Downing & Ross, 2014). Despite these theoretical concerns, observed instances of patients independently modifying treatment are rare (Carere et al., 2017). Regardless, it is a common practice for patient education materials to reiterate that modifying therapy should only be done with provider approval and oversight.

## Right CDS intervention format

CDS can be displayed in a variety of formats, such as an alert, an order set, or simply providing reference information (Freimuth et al., 2017;

Osheroff et al., 2009). These options vary on how disruptive they are in the workflow, and where in the workflow they are deployed. The selection of an appropriate CDS format is multifaceted and depends on the target end user, the local workflow, how critical the information being conveyed is, and software capabilities.

Interruptive alerts require acknowledgment to continue the workflow. Typically, these alerts "pop-up" in the foreground of the EHR and prevent the ordering process from continuing until acknowledged. Generally speaking, interruptive alerts should be reserved for highly actionable scenarios in which the risk and severity of an adverse outcome are high. One should consider strategies to reduce the frequency at which interruptive alerts fire to avoid alert fatigue. For example, if an end user has already acknowledged an alert for a patient, a second alert could be suppressed if the same user prescribes the drug again for the same patient shortly thereafter (Liu et al., 2021).

In most medication ordering scenarios, the goal of the CDS should be to guide, not interrupt the prescribing process. Ideally, the prescriber should have access to relevant pharmacogenomic information before submitting the order. This allows the prescriber to make an informed decision initially, instead of attempting to prescribe a medication and hoping that an alert is not triggered (Liu et al., 2021). Noninterruptive warnings are more passive than interruptive alerts, as they do not require acknowledgment to continue the workflow. For pharmacogenomics, this type of CDS has been deployed as information displayed inline during the medication ordering process or as part of an order set. General informational warnings or patient-specific recommendations for alternative treatment can be conveyed in this way (Aquilante et al., 2020).

InfoButtons, or hyperlinks to external resources, are a more passive approach in which the clinician may or may not choose to seek out the information. Examples in clinical pharmacogenomic include hyperlinks to more detailed, nonessential information that supplement interruptive alerts. Specifically, an interruptive alert may include an interpretation and treatment recommendation accompanied by a hyperlink to a guideline, which provides more detail about the genotype/phenotype association, population allele frequencies, supporting primary literature, and strength of recommendation. Alternatively, hyperlinks can steer end users to internal knowledge-base resources, if preferred. InfoButtons can also be deployed outside of alert text, providing patient-specific information in various navigation sections of the EHR (Aquilante et al., 2020).

Dosing calculators and algorithms have been deployed to enable clinical pharmacogenomics (Caraballo et al., 2017; Obeng et al., 2016).

Pharmacogenomic-guided dosing of warfarin is one such example. Warfarin dose requirement is a polygenic (dependent on multiple genes) phenotype that is also highly influenced by clinical and demographic variables (Johnson et al., 2017; Millican et al., 2007). Algorithms have been implemented to determine whether pharmacogenomic-guided dosing or "traditional" dosing is appropriate based on self-identified ancestry and the presence or absence of other requisite predictors of dose in the EHR (Obeng et al., 2016).

Incorporating pharmacogenomics into order sets is another CDS modality that some institutions have employed. *CYP2C19* genotype for the selection of an antiplatelet medication is believed to be most relevant in the setting of percutaneous coronary intervention (Scott et al., 2013). Early adopters of *CYP2C19*-guided antiplatelet selection reported incorporating the pharmacogenomic test order into the percutaneous coronary intervention order set to encourage use of the test in this context while minimizing irrelevant or nonevidence based CDS in the context of antiplatelet treatment for other indications (Empey et al., 2018).

Delivering messages to clinicians within the inbox functionality of the EHR has been reported by some institutions implementing pharmacogenomics (Caraballo et al., 2017). This modality may be useful for conveying information to a prescriber when the treatment initiation precedes the pharmacogenomic test result. In such a setting, interruptive or noninterruptive alerts during the ordering process are unsuitable due to lack of a test result. This may be particularly relevant when testing is ordered in the inpatient setting and the patient is discharged before test results are available. Alternatively, inbox messages may notify prescribers that a pharmacogenomic test ordered by another provider may affect that treatment. For example, a patient treated with the antidepressant paroxetine may undergo an orthopedic procedure, during which the orthopedist orders a *CYP2D6* genotype to inform analgesia selection. The paroxetine prescriber could then receive the *CYP2D6* result in his/her inbox with an interpretation or recommendation for the paroxetine treatment. This functionality may also be a viable solution when there is a reinterpretation of a past test result.

## Right time in the workflow

The medication management cycle includes medication selection, medication reconciliation, ordering, verification/dispensing, administration, patient education, and monitoring (Osheroff et al., 2009). Although pharmacogenomics could be applied to each of these tasks, the highest priority areas for using pharmacogenomics to improve medication safety and effectiveness are selection, ordering, and verification (Bell et al., 2014).

The "right time" for CDS in the workflow can be before or after pharmacogenomic results are available. In the context of reactive testing, test turnaround time usually exceeds clinical decision-making time (Freimuth et al., 2017). For example, CDS may be used to remind clinicians that pharmacogenomic testing is required or recommended prior to initiating specific treatments (e.g., testing for *HLA-B**57:01 prior to initiating abacavir) (Bell et al., 2014; Caraballo et al., 2017). Likewise, the option to order pharmacogenomic testing can be integrated into order sets for different clinical scenarios.

Identifying the "right time" requires a strong understanding of local workflow. It may differ from clinic to clinic within a health system. The ideal time to order and collect a specimen for pharmacogenomic testing likely depends on local workflow. Can an order for pharmacogenomic testing precede treatment initiation—for example, in the setting of elective surgery? Will ordering testing prior preemptively to guide treatment initiation present reimbursement issues? Is the patient providing blood samples for other laboratory tests during the routine workflow? These detailed questions demonstrate why it is so important to engage a wide variety of administrative stakeholders and end users during the design and modification of CDS.

## Right channel

Generally speaking, CDS can be applied during any part of the medication management cycle. While the EHR is ripe with opportunity for CDS, numerous outside technologies can house CDS tools as well. For example, medication dispensing cabinets, barcoding software, and intravenous infusion pumps are all valuable technologies that can use CDS to improve medication administration safety, but these channels are less often used to implement pharmacogenomics.

While some institutions have reported success in using native EHR functionality for pharmacogenomics implementation (Aquilante et al., 2020), it should be mentioned that all EHRs are not necessarily built or equipped to handle genetic or genomic information. Some institutions have turned to third-party options such as an ancillary omic system. An ancillary omic system allows for storage of large genomic data files, in addition to tools and processing pipelines that enable the clinical interpretation of these data (Rasmussen et al., 2019).

# Implementing and maintaining manually developed rules

## Pharmacogenomic governance for CDS

Many institutions customize the knowledge base used to inform the local drug–drug interaction CDS (Osheroff et al., 2009). In the same way, local customization is common in clinical pharmacogenomics. Given the laboratory, informatics, and clinician education requirements of implementing pharmacogenomics, many institutions choose to implement one or a few drug–gene interactions at a time. Stakeholders must identify the local patient populations, clinics, and provider groups that present the best opportunity for pharmacogenomics to improve treatment outcomes and efficiency. Many institutions create a precision medicine team or service to lead these efforts (Duarte et al., 2021). These teams are almost always multidisciplinary, consisting of physicians, pharmacists, pharmacologists, pathologists, geneticists, health informaticists, and others (Bell et al., 2014; Duarte et al., 2021; Liu et al., 2021). Because many aspects of drug–gene interaction CDS are similar to drug–drug interaction CDS, it may be possible to use established informatics personnel and processes to streamline development, testing, implementation, and maintenance of pharmacogenomic CDS (Caraballo et al., 2017). Likewise, leaning on established sources of local guidance, such as the Pharmacy and Therapeutics committee, may give end users greater confidence in the clinical utility of the recommendations (Bell et al., 2014; Liu et al., 2021).

## Sources of pharmacogenomic guidance—the need for literature review

Although treatment recommendations based on genotype are available from the FDA and consortia such as the Clinical Pharmacogenetics Implementation Consortium (CPIC), it is up to the healthcare system to determine how critical these recommendations are for prescribers and how to efficiently add CDS to local workflows. Most clinicians would agree that the pharmacogenomic recommendation to start the selective serotonin reuptake inhibitor sertraline at a reduced dose is less critical than warning that the patient is at high risk for a potentially fatal hypersensitivity reaction to abacavir. Indeed, the corresponding information in FDA-approved labels for these pharmacogenomic recommendations vary from not mentioned (reducing sertraline starting dose in *CYP2C19* poor metabolizers) to a black box warning (initiating abacavir in an *HLA-B**57:01 carrier). Likewise, CPIC classifies the strength of the therapeutic recommendation as strong, moderate, or optional based upon a combination of the evidence quality and the advantages and disadvantages of modifying therapy (Caudle et al., 2016). However, it is left to leadership in each health system to determine the appropriate verbiage and CDS modality that combines the classification of recommendation, severity of the outcome, frequency that the medication is prescribed within the system, and local workflow.

In addition, there is no singular resource to provide all the sufficient information to build CDS for all actionable pharmacogenomic drug–gene pairs. There are drug–gene interactions that do not have CPIC guidelines, but have prescribing recommendations from other consortia (e.g., the Dutch Pharmacogenetics Working Group [DPWG]), clinical practice guidelines (e.g., the National Comprehensive Cancer Network), or FDA-approved labeling. The FDA-approved labeling of the multiple sclerosis medication siponimod instructs prescribers to determine *CYP2C9* genotype prior to initiating treatment. The starting dose and whether the drug is contraindicated depends on genotype. However, CPIC guidelines are not available for this somewhat rarely used medication.

Furthermore, recommendations from different resources can be conflicting. Nearly half of actionable drug–gene interactions have inconsistent guidance regarding recommendation category, whether routine testing is recommended, and patient subpopulations requiring treatment modification (Shugg et al., 2020). Much in the same way that consortia and clinical practice guidelines may reach different conclusions based on available evidence, local clinical groups and stakeholders may reach a different conclusion than guideline authors. For example, some institutions have implemented CDS for some, but not all of the selective serotonin reuptake inhibitors included in CPIC guidelines, based on local interpretation of strength of evidence (Aquilante et al., 2020). In totality, these issues demonstrate the need for local literature review rather than simply relying on a single external resource to determine actionable drug–gene interactions.

## Implementing a new drug–gene interaction

Selecting drug–gene interactions for local implementation requires consideration of (1) the institution's patient population, (2) clinical needs of patients and providers, and (3) programmatic resources and infrastructure

(Liu et al., 2021). Considerations around the patient population include what disease states are treated at the institution, the health insurance status and/or socioeconomic status of patients, and actionable phenotype. The clinical needs of patients and providers may inform how detailed the information provided by CDS needs to be, or how direct the treatment recommendations are. Likewise, institutional clinical needs may be influenced by the health system formulary or local prescribing patterns—medications that are rarely prescribed will generally be lower priority. Programmatic resources and infrastructure may dictate what channels are viable for communicating with clinicians and patients, whether testing is done "in-house" or sent to a third-party laboratory, and where and how genomic data are stored. Furthermore, the strength of evidence, volume of prescribing, breadth of prescribing, and anticipated reimbursement are critical to selecting drug–gene interaction for implementation (Caraballo et al., 2017).

Another consideration is which data are going to trigger the CDS? Will the triggering criteria be genotype, phenotype, activity score, or something else? Developing CDS based off a phenotype is likely less work up front but may present complications down the road (for example, if the genotype-to-phenotype translation changes) whereas building CDS based off of genotype requires more time prior to implementation but may save time in the long run. Table 8.1 provides an example of a warning based upon phenotype that can be built utilizing Boolean logic.

## Implementing new drugs for already tested genes

There are advantages and potential obstacles for implementing CDS for a second medication based on an already implemented pharmacogenomic test. For one, implementing CDS for all medications with actionable recommendations related to a single gene's genotype reduces the probability that a provider will prescribe a medication without being aware that relevant pharmacogenomic results exist—a potential liability situation. In addition, this type of programmatic expansion does not require a new laboratory assay, and at least some providers will be familiar with the pharmacogenomic importance of the genetic assay (Bell et al., 2014).

Many institutions chose *CYP2C19* genotyping for the purpose of antiplatelet selection as one of the first drug–gene pairs for implementation (Aquilante et al., 2020; Liu et al., 2021). However, *CYP2C19* genotype is relevant for the prescribing of many commonly used drugs outside of the antiplatelet class. When implementing additional drug–gene interactions

**Table 8.1** Example of a knowledge-based warning that can be built utilizing boolean logic.

| Alert section | Conditional knowledge |
|---|---|
| Conditions | CYP2C19 poor metabolizer OR intermediate metabolizer AND clopidogrel |
| Header | Clopidogrel is not recommended due to increased risk of stent thrombosis and cardiovascular events |
| Body | **Treatment modification is recommended if not contraindicated**: Prasugrel (EFFIENT) 10 mg daily and discontinue clopidogrel (PLAVIX)<br>Contraindications:<br>- History of stroke, TIA, or active bleeding<br>- Age greater than 75 years<br>- Weight less than 60 kg<br>- Ticagrelor (BRILINTA) 90 mg twice daily and discontinue clopidogrel (PLAVIX)<br>Contraindications:<br>- History of intracranial hemorrhage or active bleed<br>- Hepatic impairment<br>CAUTION: maintenance doses of aspirin above 100 mg reduce effectiveness of ticagrelor and should be avoided<br>**If continuing clopidogrel (PLAVIX), please acknowledge your reason below** |
| CDS options | - Ability to remove clopidogrel order<br>- Ability to order ticagrelor inpatient/outpatient<br>- Ability to order prasugrel inpatient/outpatient<br>- Acknowledge reason if clinician chose to override warning |

related to *CYP2C19*, the test interpretations displayed in the EHR may need to be changed. For example, if the first interpretation was drug-based (e.g., "clopidogrel poor metabolizer"), a new drug-based interpretation will need to be added (e.g., "citalopram poor metabolizer") or the interpretations must be shifted to gene-based (e.g., "CYP2C19 poor metabolizer").

## Reinterpretation

Although an individual's genomic sequence should not change over time, interpretations and treatment recommendations do change as new variation is identified in the gene of interest and scientific understanding of the genotype/phenotype relationship grows. The need for reinterpretation is not a theoretical possibility, but a certain reality in the rapidly expanding field of pharmacogenomics. Thus, infrastructure and governance should be built knowing that continual review and modification will be necessary.

### *Changing guidance*

By design, clinical practice guidelines, CPIC guidelines, and the FDA Table of Pharmacogenetic Associations are periodically updated. These updates can result in major changes to guidance. Recent examples of changing guidance from CPIC guideline updates include revised dose adjustment recommendations (Amstutz et al., 2018), expansion of recommendations to other medications within a drug class (Crews et al., 2021), and consideration of additional genes in clinical decision-making (Johnson et al., 2017). Intuitively, components of pharmacogenomic CDS will need to be revised to maintain consistency with this source material.

However, guideline revisions require a literature search and establishment of consensus from expert authors. Because of this process, guideline revisions can lag several months or years after landmark studies are released. This is an additional reason to plan for regular literature review and discussion with stakeholders at the health system level.

### *Changing nomenclature and genotype/phenotype translation*

Standardized nomenclature is important to allow for the sharing of pharmacogenetic results across laboratories, EHRs, and health systems (Caudle et al., 2017). Nevertheless, this standardized nomenclature may change as our understanding of genotype/phenotype relationships and expert opinions evolve. The term, "extensive metabolizer," generally referring to individuals with two functional alleles for a drug metabolizing enzyme, was replaced with "normal metabolizer" in CPIC guidelines as some clinicians were often confused by the older term (Caudle et al., 2017). Additionally, expert opinion on how to translate genotype to predicted phenotype may change. CPIC and DPWG recently modified the standardized process for assigning *CYP2D6* predicted phenotype based on genotype (Caudle et al., 2020). This standardization effort resulted in both organizations changing their recommendations for interpreting phenotype from genotype. Institutions that utilized phenotype for CDS were then faced with the dilemma of what they should do with historical results; should they all be updated based on the new recommendations? Should they be left as is, and if so, how does this affect CDS going forward? How should patients and providers be notified?

Complicating matters further, updated recommendations for certain medication classes are based on the activity score instead of phenotype. Confusingly, this can result in different treatment recommendations for patients with the same predicted phenotype classification. For example, current CPIC recommendations for phenytoin include a reduction of starting dose for individuals with a *CYP2C9* activity score of 1, but no dosage adjustment is recommended for individuals with an activity score of 1.5 even though both are classified as intermediate *CYP2C9* metabolizers (Karnes et al., 2021). Previous phenytoin guidelines provided dosing recommendations solely based on phenotype (Caudle et al., 2014). Table 8.2 illustrates how an individual with the *CYP2C9* genotype *1/*2 would have different dosing recommendations depending on when the interpretation was conducted.

### *Considering result format in the EHR and communication of reinterpretations*

When planning a workflow for reinterpreting results, one should consider the format in which results are displayed within the EHR. For example, PDF results cannot be easily updated since they are static. Optimal methods for notifying providers of reinterpreted results also needs to be considered. The urgency for notifying providers and/or patients of reinterpretations should be dictated by difference of actionability of the interpretation. For example, if a treatment recommendation for a given patient has changed due to reinterpretation, this is a more urgent message to convey than a reinterpretation that affects the patient's activity score but not the predicted phenotype or treatment recommendation for actively prescribed medications. Establishing a standard communication channel and template for communicating reinterpretations proactively can ease the reinterpretation process.

## Limitations of manually developed rules

Conceptually, applying precision medicine to an individual should include consideration of all patient-specific factors into prescribing—age, gender, race, renal function, insurance status, etc. However, most pharmacogenomic CDS uses genotype and medication orders without considering these other patient-specific factors. That is, CDS rules can often ignore nongenetic factors that intuitively override a treatment recommendation or recommendation to order a pharmacogenomic test on the patient. More advanced CDS may include multiple sets of logic not only to encompass drug—gene interactions but also include factors such as disease state, organ function, etc. For instance, the earlier example of clopidogrel-*CYP2C19* drug—gene interaction alert could further be refined to exclude the alternative medication prasugrel if the patient has a history of stroke or transient ischemic attack (contraindications to prasugrel treatment).

**Table 8.2** Example of changing treatment recommendations for a patient with *CYP2C9* *1/*2 genotype.

| Guideline | *CYP2C9* genotype | Phenotype | Activity score | Therapeutic recommendation |
|---|---|---|---|---|
| 2014 CPIC Phenytoin Guidelines (Caudle et al., 2014) | *1/*2 | Intermediate metabolizer | N/A | Consider 25% reduction of recommended starting maintenance dose. Subsequent maintenance doses should be adjusted according to therapeutic drug monitoring and response. |
| 2020 CPIC Phenytoin Guidelines (Karnes et al., 2021) | *1/*2 | Intermediate metabolizer | 1.5 | No adjustments needed from typical dosing strategies. Subsequent doses should be adjusted according to therapeutic drug monitoring, response, and side effects. |

Table is for illustrative purposes only. For full and current recommendations refer to the most updated CPIC guidelines.

There are clinical situations in which the genetically predicted phenotype does not match the observed phenotype, termed phenoconversion (Shah & Smith, 2015). As an extreme example, genotype obtained from a peripheral blood sample in a patient who has received a liver transplant may incorrectly characterize the metabolic activity of drug metabolizing enzymes (i.e., cytochrome P450 enzymes) from the transplanted liver (Provenzani et al., 2013). A more common type of phenoconversion is the result of a drug–drug interaction (Fig. 8.1). For example, a patient genetically predicted to be a *CYP2D6* normal metabolizer may exhibit the characteristics of a CYP2D6 poor metabolizer (e.g., high parent drug exposure) in the presence of a strong inhibitor of CYP2D6 (e.g., paroxetine). In such patients, CDS that identifies a drug–gene interaction but ignores the possibility of drug-induced phenoconversion may provide incorrect treatment recommendations. The relevance of phenoconversion for pharmacogenomic-guided treatment recommendations is now being recognized in CPIC guidelines (Crews et al., 2021).

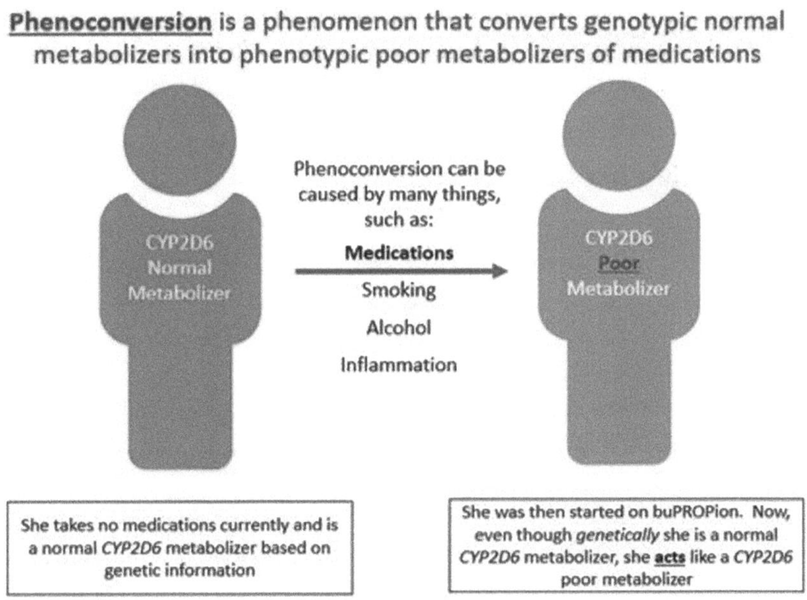


**Figure 8.1** Example of drug-induced phenoconversion. Phenoconversion can occur when a drug causes such complete inhibition of a metabolizing enzyme that a genotypic normal metabolizer acts like (phenotype is converted to) a poor metabolizer. This interaction is reversible upon discontinuation of the inhibiting drug. *Modified with permission from (Petry & Van Heukelom, 2020).*

There are also technical limitations of manually developed CDS rules. For one, unstructured results (e.g., a PDF) typically cannot be used to fire result-specific CDS within the EHR, therefore preventing the availability of clinical guidance upon order entry. Although many institutions are working to standardize the way that pharmacogenomic tests are ordered and resulted, historic results may be found in many different areas of the EHR—the problem list, allergy section, laboratory results, genomic indicators, or as nonstructured media (e.g., a PDF) (Caraballo et al., 2020). Thus, pharmacogenomic CDS may need to include potential firing criteria from any of these areas. Alternatively, or as a complimentary process, a manual review of historical results with documentation in the proper area of the EHR can be employed.

In addition, some variables that should be used in clinical decision-making are not well- or easily captured in the EHR. Many pharmacogenomic treatment recommendations involve reducing starting doses and are not applicable on patients on long-term treatment that have already been titrated to an appropriate dose. The automated detection of long-term therapy versus new treatment initiations can be difficult if home medication regimens are not accurately and discretely characterized. Similarly, patient ancestry can be relevant for pharmacogenomic decision making, as some clinically important alleles are only found, or only found at high frequency in certain ethnic groups.

## Predictive algorithms in pharmacogenomics

While many pharmacogenomic CDS rules are relatively simple single gene scenarios—if predicted phenotype is X then recommend Y—other known pharmacogenomic associations are more complex. This lends itself to the clinical use of algorithms in pharmacogenomics. These algorithms often use both genetic and nongenetic data as inputs.

Warfarin dose requirement is known to be a polygenic trait that is also strongly influenced by nongenetic factors (e.g., drug interactions, diet, age, etc.). There are a variety of algorithms that have been developed to predict warfarin dose requirement (Tavares et al., 2018). The CPIC guideline recommendations for genotype-guided warfarin dosing were changed to a decision tree as of 2017, with branches related to which alleles were tested and the self-identified race of the patient (Johnson et al., 2017). Building CDS for dosing algorithms like these can be difficult, as some of the important nongenetic predictors can be missing or inaccurate in the EHR (e.g., race, smoking status).

Algorithms have also been developed for selecting psychiatric medications, most commonly antidepressants. These tools, sometimes called "combinatorial" pharmacogenomics, use genotypes or phenotypes from multiple pharmacokinetic (*CYP2D6, CYP2C19, CYP2C9, CYP1A2*) and pharmacodynamic (*SLC6A4, HTR2A*) genes together to help clinicians select an antidepressant agent (Altar et al., 2015). While some positive outcomes have been reported (Altar et al., 2015; Rothschild et al., 2021; Winner et al., 2015), the proprietary nature of the algorithms, inconsistent reimbursement, and difficult integrating the results with the EHR remain barriers to widespread adoption. It is important to remember that genes assayed for combinatorial algorithms are individually important for nonpsychiatric drugs (e.g., *CYP2D6* and codeine)—so discrete results in the EHR enabling CDS are preferable to the paper or PDF reports typically provided by these products.

Outside of drug selection and dosing, algorithms have been used to identify patients likely to benefit from pharmacogenomic testing. These algorithms, while clearly related to pharmacogenomic, do not use any patient-level genetic information. Instead, pharmacogenomic allele frequencies are cross-referenced with the patient's medication list to predict the likelihood that running a multigene pharmacogenomic panel would yield an actionable pharmacogenomic recommendation (Miller, 2019). Although some of these analyses have been done with proprietary algorithms (Heise et al., 2020), others have published analyses using more transparent methods to estimate the clinical utility of pharmacogenomics in patients with specific diagnoses or undergoing specific procedures (Black et al., 2020; El Rouby et al., 2020; Stevenson et al., 2021). These types of prediction tools may become more important as financial reimbursement rates for multigene panels increase. Finally, some proprietary tools use pharmacogenomic results as an input for predicting risk of adverse drug reactions (Bankes et al., 2020). Such tools can enable risk stratification and more appropriate resource allocation (e.g., more regular monitoring or medication management visits) for patients at higher risk for adverse drug reactions.

## Clinical considerations for healthcare disparities

The term healthcare disparities refers to systematic differences in access or quality of care between groups, driven by underlying social advantage or social hierarchies. For example, access or quality of healthcare may be reduced for women, ethnic minorities, people living in rural communities, or low-income individuals. There are several mechanisms by which pharmacogenomics could either widen or reduce healthcare disparities (Martin et al., 2014).

There are several considerations related to race, ethnicity, and/or ancestry that are notable in translating pharmacogenomic CDS into clinical practice. For one, certain important pharmacogenomic alleles are only found at appreciable frequency in individuals with a certain ethnic background. For example, the *HLA-B**15:02 allele predicts excessive risk of Stevens–Johnson Syndrome during carbamazepine treatment and is much more common in patients with Chinese ancestry than European or African ancestry. As a result, the FDA-approved labeling recommends screening for this allele in patients of Chinese ancestry. The EHR's discrete labels for patient race are likely inadequate for identifying patients that may carry ethnicity-specific alleles. In practice, a patient with a father of Chinese ancestry and a mother of African ancestry may have their race listed as "Asian," "Black," or "Other" in the EHR, with a CDS alert recommending pretreatment testing not firing in the latter two scenarios. In addition, broad terms like "Asian or Pacific Islander" or "Black" can be too heterogeneous for pharmacogenomic purposes. Subgroups within these broad terms like the Hmong (an Asian subpopulation) have allele frequencies that significantly differ from the allele frequencies reported in more general East Asian populations (Sun et al., 2021).

Similarly, the "wild-type" or most common phenotype for a pharmacogenomic gene can differ by ancestry. For the gene *CYP3A5*, "poor metabolizer" is the most common phenotype in European ancestry populations, while "normal metabolizer" is the most common phenotype for African ancestry populations (Birdwell et al., 2015). Thus, a blanket statement such as "we should implement interruptive CDS for all poor metabolizers" should be met with skepticism—such interruptive alerts would fire for most European ancestry patients, potentially contributing to alert fatigue.

Concerns about clinical pharmacogenomics widening disparities are not purely hypothetical. In the Clarification of Oral Anticoagulation through Genetics (COAG) trial, researchers intended to improve dosing of warfarin by incorporating pharmacogenomics in addition to nongenetic factors (Kimmel et al., 2013). In participants of African ancestry, however, the incorporation of pharmacogenomics performed *worse* than nongenetic factors alone. The algorithm used in the trial was built from research in primarily European ancestry patients, was known to perform worse in African individuals versus European individuals, and did not include African-specific alleles (Oni-Orisan et al., 2021). Pharmacogenomic implementers should consider the populations from which pharmacogenomic discoveries are identified and whether there are external validity issues in respect to their local patient population before incorporating pharmacogenomics into clinical care.

At the same time, pharmacogenomics could be a key piece to reducing healthcare disparities related to medication outcomes. The discovery of ancestry-specific alleles that increase risk of treatment failure or adverse drug reactions may partially explain known disparities in treatment outcomes and enable personalized treatment plans that produce more equitable outcomes. A greater focus on performing research in which participants of all types are represented and thoughtful implementation into care can help us achieve the ultimate goal of precision medicine: tailored treatment that improves health outcomes for all.

## Conclusion

This chapter discussed essential considerations for effective CDS. Despite the availability of clinical practice guidelines and increasing standardization, local governance and evidence review is required to meet the needs of a specific healthcare system. Including clinical leaders and end users in the iterative process of designing and refining precision medicine CDS is recommended to best integrate CDS into clinical workflows and increase provider acceptance. At the same time, several challenges and limitations were acknowledged, including the certainty that precision medicine CDS will need to be periodically revised and concerns that pharmacogenomics could worsen healthcare disparities. Recognizing these best practices and creating processes to overcome anticipated challenges can lead to a successful clinical implementation of precision medicine CDS.

## References

Altar, C. A., Carhart, J. M., Allen, J. D., Hall-Flavin, D. K., Dechairo, B. M., & Winner, J. G. (2015). Clinical validity: Combinatorial pharmacogenomics predicts antidepressant responses and healthcare utilizations better than single gene phenotypes. *Pharmacogenomics Journal, 15*(5), 443–451. https://doi.org/10.1038/tpj.2014.85

Amstutz, U., Henricks, L. M., Offer, S. M., Barbarino, J., Schellens, J. H. M., Swen, J. J., Klein, T. E., McLeod, H. L., Caudle, K. E., Diasio, R. B., & Schwab, M. (2018). Clinical pharmacogenetics implementation consortium (CPIC) guideline for dihydropyrimidine dehydrogenase genotype and fluoropyrimidine dosing: 2017 update. *Clinical Pharmacology and Therapeutics, 103*(2), 210–216. https://doi.org/10.1002/cpt.911

Aquilante, C. L., Kao, D. P., Trinkley, K. E., Lin, C. T., Crooks, K. R., Hearst, E. C., Hess, S. J., Kudron, E. L., Lee, Y. M., Liko, I., Lowery, J., Mathias, R. A., Monte, A. A., Rafaels, N., Rioth, M. J., Roberts, E. R., Taylor, M. R. G., Williamson, C., & Barnes, K. C. (2020). Clinical implementation of pharmacogenomics via a health system-wide research biobank: The University of Colorado experience. *Pharmacogenomics, 21*(6), 375–386. https://doi.org/10.2217/pgs-2020-0007

Bankes, D., Jin, H., Finnel, S., Michaud, V., Knowlton, C., Turgeon, J., & Stein, A. (2020). Association of a novel medication risk score with adverse drug events and other pertinent outcomes among participants of the programs of all-inclusive care for the elderly. *Pharmacy, 8*(2), 87. https://doi.org/10.3390/pharmacy8020087

Beitelshees, A. L., Stevenson, J. M., El Rouby, N., Dillon, C., Empey, P. E., Fielstein, E. M., Johnson, J. A., Limdi, N. A., Ong, H. H., Franchi, F., Angiolillo, D. J., Peterson, J. F., Rosenman, M. B., Skaar, T. C., Tuteja, S., & Cavallari, L. H. (2020). Evaluating the extent of reusability of CYP2C19 genotype data among patients genotyped for antiplatelet therapy selection. *Genetics in Medicine, 22*(11), 1898–1902. https://doi.org/10.1038/s41436-020-0894-2

Bell, G. C., Crews, K. R., Wilkinson, M. R., Haidar, C. E., Hicks, J. K., Baker, D. K., Kornegay, N. M., Yang, W., Cross, S. J., Howard, S. C., Freimuth, R. R., Evans, W. E., Broeckel, U., Relling, M. V., & Hoffman, J. M. (2014). Development and use of active clinical decision support for preemptive pharmacogenomics. *Journal of the American Medical Informatics Association, 21*(2), e93–e99. https://doi.org/10.1136/amiajnl-2013-001993

Birdwell, K., Decker, B., Barbarino, J., Peterson, J., Stein, C., Sadee, W., Wang, D., Vinks, A., He, Y., Swen, J., Leeder, J., van Schaik, R., Thummel, K., Klein, T., Caudle, K., & MacPhee, I. (2015). Clinical Pharmacogenetics Implementation Consortium (CPIC) guidelines for CYP3A5 genotype and tacrolimus dosing. *Clinical Pharmacology & Therapeutics, 98*(1), 19–24. https://doi.org/10.1002/cpt.113

Black, R. M., Williams, A. K., Ratner, L., Crona, D. J., Wiltshire, T., Weck, K. E., Stouffer, G. A., & Lee, C. R. (2020). Projected impact of pharmacogenomic testing on medications beyond antiplatelet therapy in percutaneous coronary intervention patients. *Pharmacogenomics, 21*(7), 431–441. https://doi.org/10.2217/pgs-2019-0185

Blagec, K., Koopmann, R., Crommentuijn-Van Rhenen, M., Holsappel, I., Van Der Wouden, C. H., Konta, L., Xu, H., Steinberger, D., Just, E., Swen, J. J., Guchelaar, H. J., & Samwald, M. (2018). Implementing pharmacogenomics decision support across seven European countries: The Ubiquitous Pharmacogenomics (U-PGx) project. *Journal of the American Medical Informatics Association, 25*(7), 893–898. https://doi.org/10.1093/jamia/ocy005

Blagec, K., Romagnoli, K. M., Boyce, R. D., & Samwald, M. (2016). Examining perceptions of the usefulness and usability of a mobile-based system for pharmacogenomics clinical decision support: A mixed methods study. *PeerJ, 2016*(2), e1671. https://doi.org/10.7717/peerj.1671

Caraballo, P. J., Hodge, L. S., Bielinski, S. J., Stewart, A. K., Farrugia, G., Schultz, C. G., Rohrer-Vitek, C. R., Olson, J. E., St Sauver, J. L., Roger, V. L., Parkulo, M. A., Kullo, I. J., Nicholson, W. T., Elliott, M. A., Black, J. L., & Weinshilboum, R. M. (2017). Multidisciplinary model to implement pharmacogenomics at the point of care. *Genetics in Medicine, 19*(4), 421–429. https://doi.org/10.1038/gim.2016.120

Caraballo, P. J., Sutton, J. A., Giri, J., Wright, J. A., Nicholson, W. T., Kullo, I. J., Parkulo, M. A., Bielinski, S. J., & Moyer, A. M. (2020). Integrating pharmacogenomics into the electronic health record by implementing genomic indicators. *Journal of the American Medical Informatics Association, 27*(1), 154–158. https://doi.org/10.1093/jamia/ocz177

Carere, D. A., Vanderweele, T. J., Vassy, J. L., Van Der Wouden, C. H., Roberts, J. S., Kraft, P., Green, R. C., Krier, J. B., Helm, M. H., Kalia, S. S., Christensen, K. D., Lehmann, L. S., Ruffin, M. T., Le, L. Q., Ostergren, J., Uhlmann, W. R., Couper, M. P., Mountain, J. L., Kiefer, A. K., ... Kaphingst, K. (2017). Prescription medication changes following direct-to-consumer personal genomic testing: Findings from the Impact of Personal Genomics (PGen) Study. *Genetics in Medicine, 19*(5), 537–545. https://doi.org/10.1038/gim.2016.141

Caudle, K. E., Dunnenberger, H. M., Freimuth, R. R., Peterson, J. F., Burlison, J. D., Whirl-Carrillo, M., Scott, S. A., Rehm, H. L., Williams, M. S., Klein, T. E., Relling, M. V., & Hoffman, J. M. (2017). Standardizing terms for clinical pharmacogenetic test results: Consensus terms from the Clinical Pharmacogenetics Implementation Consortium (CPIC). *Genetics in Medicine, 19*(2), 215–223. https://doi.org/10.1038/gim.2016.87

Caudle, K. E., Gammal, R. S., Whirl-Carrillo, M., Hoffman, J. M., Relling, M. V., & Klein, T. E. (2016). Evidence and resources to implement pharmacogenetic knowledge for precision medicine. *American Journal of Health-System Pharmacy, 73*(23), 1977–1985. https://doi.org/10.2146/ajhp150977

Caudle, K. E., Rettie, A. E., Whirl-Carrillo, M., Smith, L. H., Mintzer, S., Lee, M. T. M., Klein, T. E., & Callaghan, J. T. (2014). Clinical pharmacogenetics implementation consortium guidelines for CYP2C9 and HLA-B genotypes and phenytoin dosing. *Clinical Pharmacology & Therapeutics, 96*(5), 542–548. https://doi.org/10.1038/clpt.2014.159

Caudle, K. E., Sangkuhl, K., Whirl-Carrillo, M., Swen, J. J., Haidar, C. E., Klein, T. E., Gammal, R. S., Relling, M. V., Scott, S. A., Hertz, D. L., Guchelaar, H. J., & Gaedigk, A. (2020). Standardizing CYP2D6 genotype to phenotype translation: Consensus recommendations from the clinical pharmacogenetics implementation consortium and Dutch pharmacogenetics working group. *Clinical and Translational Science, 13*(1), 116–124. https://doi.org/10.1111/cts.12692

Crews, K., Monte, A., Huddart, R., Caudle, K., Kharasch, E., Gaedigk, et al. (2021). Clinical pharmacogenetics implementation consortium guideline for CYP2D6, OPRM1, and COMT genotypes and select opioid therapy. *Clinical Pharmacology & Therapeutics, 110.*

Downing, N. S., & Ross, J. S. (2014). Innovation, risk, and patient empowerment the FDA-mandated withdrawal of 23andMe's personal genome service. *JAMA - Journal of the American Medical Association, 311*(8), 793–794. https://doi.org/10.1001/jama.2014.148

Duarte, J. D., Dalton, R., Elchynski, A. L., Smith, D. M., Cicali, E. J., Lee, J. C., Duong, B. Q., Petry, N. J., Aquilante, C. L., Beitelshees, A. L., Empey, P. E., Johnson, J. A., Obeng, A. O., Pasternak, A. L., Pratt, V. M., Ramsey, L. B., Tuteja, S., Van Driest, S. L., Wiisanen, K., ... Cavallari, L. H. (2021). Multisite investigation of strategies for the clinical implementation of pre-emptive pharmacogenetic testing. *Genetics in Medicine.* https://doi.org/10.1038/s41436-021-01269-9, 34282303.

El Rouby, N., Alrwisan, A., Langaee, T., Lipori, G., Angiolillo, D. J., Franchi, F., Riva, A., Elsey, A., Johnson, J. A., Cavallari, L. H., & Winterstein, A. G. (2020). Clinical utility of pharmacogene panel-based testing in patients undergoing percutaneous coronary intervention. *Clinical and Translational Science, 13*(3), 473–481. https://doi.org/10.1111/cts.12729

Empey, P. E., Stevenson, J. M., Tuteja, S., Weitzel, K. W., Angiolillo, D. J., Beitelshees, A. L., Coons, J. C., Duarte, J. D., Franchi, F., Jeng, L. J. B., Johnson, J. A., Kreutz, R. P., Limdi, N. A., Maloney, K. A., Owusu Obeng, A., Peterson, J. F., Petry, N., Pratt, V. M., Rollini, F., ... Lee, C. R. (2018). Multisite investigation of strategies for the implementation of CYP2C19 genotype-guided antiplatelet therapy. *Clinical Pharmacology and Therapeutics, 104*(4), 664–674. https://doi.org/10.1002/cpt.1006, 29280137.

Freimuth, R., Formea, C., Hoffman, J., Matey, E., Peterson, J., & Boyce, R. (2017). Implementing genomic clinical decision support for drug-based precision medicine. *CPT: Pharmacometrics & Systems Pharmacology, 6*(3), 153–155. https://doi.org/10.1002/psp4.12173

Friedberg, M. W., Chen, P. G., Van Busum, K. R., Aunon, F., Pham, C., Caloyeras, J., Mattke, S., Pitchforth, E., Quigley, D. D., Brook, R. H., Crosson, F. J., & Tutty, M. (2014). Factors affecting physician professional satisfaction and their implications for patient care, health systems, and health policy. *RAND Health Quarterly, 3*(4).

Gage, B. F., Eby, C., Johnson, J. A., Deych, E., Rieder, M. J., Ridker, P. M., Milligan, P. E., Grice, G., Lenzini, P., Rettie, A. E., Aquilante, C. L., Grosso, L., Marsh, S., Langaee, T., Farnett, L. E., Voora, D., Veenstra, D. L., Glynn, R. J., Barrett, A., & McLeod, H. L. (2008). Use of pharmacogenetic and clinical factors to predict the therapeutic dose of warfarin. *Clinical Pharmacology and Therapeutics, 84*(3), 326–331. https://doi.org/10.1038/clpt.2008.10

Heise, C. W., Gallo, T., Curry, S. C., & Woosley, R. L. (2020). Identification of populations likely to benefit from pharmacogenomic testing. *Pharmacogenetics and Genomics*, 91–95. https://doi.org/10.1097/FPC.0000000000000400

Johnson, J. A., Caudle, K. E., Gong, L., Whirl-Carrillo, M., Stein, C. M., Scott, S. A., Lee, M. T., Gage, B. F., Kimmel, S. E., Perera, M. A., Anderson, J. L., Pirmohamed, M., Klein, T. E., Limdi, N. A., Cavallari, L. H., & Wadelius, M. (2017). Clinical Pharmacogenetics Implementation Consortium (CPIC) guideline for pharmacogenetics-guided warfarin dosing: 2017 update. *Clinical Pharmacology and Therapeutics, 102*(3), 397–404. https://doi.org/10.1002/cpt.668

Karnes, J. H., Rettie, A. E., Somogyi, A. A., Huddart, R., Fohner, A. E., Formea, C. M., Ta Michael Lee, M., Llerena, A., Whirl-Carrillo, M., Klein, T. E., Phillips, E. J., Mintzer, S., Gaedigk, A., Caudle, K. E., & Callaghan, J. T. (2021). Clinical pharmacogenetics implementation consortium (CPIC) guideline for CYP2C9 and HLA-B genotypes and phenytoin dosing: 2020 update. *Clinical Pharmacology and Therapeutics, 109*(2), 302–309. https://doi.org/10.1002/cpt.2008

Kimmel, S. E., French, B., Kasner, S. E., Johnson, J. A., Anderson, J. L., Gage, B. F., Rosenberg, Y. D., Eby, C. S., Madigan, R. A., McBane, R. B., Abdel-Rahman, S. Z., Stevens, S. M., Yale, S., Mohler, E. R., Fang, M. C., Shah, V., Horenstein, R. B., Limdi, N. A., Muldowney, J. A. S., … Ellenberg, J. H. (2013). A pharmacogenetic versus a clinical algorithm for warfarin dosing. *New England Journal of Medicine, 369*(24), 2283–2293. https://doi.org/10.1056/NEJMoa1310669. PMID: 24251361.

Klein, T. E., Altman, R. B., Eriksson, N., Gage, B. F., Kimmel, S. E., Lee, M.-T. M., Limdi, N. A., Page, D., Roden, D. M., Wagner, M. J., Caldwell, M. D., & Johnson, J. A. (2009). Estimation of the warfarin dose with clinical and pharmacogenetic data. *The New England Journal of Medicine, 360*(8), 753–764. https://doi.org/10.1056/NEJMoa0809329

Liu, M., Vnencak-Jones, C. L., Roland, B. P., Gatto, C. L., Mathe, J. L., Just, S. L., Peterson, J. F., Van Driest, S. L., & Weitkamp, A. O. (2021). A tutorial for pharmacogenomics implementation through end-to-end clinical decision support based on ten years of experience from PREDICT. *Clinical Pharmacology and Therapeutics, 109*(1), 101–115. https://doi.org/10.1002/cpt.2079

Martin, M. A., Hoffman, J. M., Freimuth, R. R., Klein, T. E., Dong, B. J., Pirmohamed, M., Hicks, J. K., Wilkinson, M. R., Haas, D. W., & Kroetz, D. L. (2014). Clinical pharmacogenetics implementation consortium guidelines for HLA-B genotype and abacavir dosing: 2014 update. *Clinical Pharmacology & Therapeutics, 95*(5), 499–500. https://doi.org/10.1038/clpt.2014.38

Miller, K. (April 2019). *Big data and clinical genomics*. Clinical OMICs. https://www.clinicalomics.com/magazine-editions/volume-6-issue-number-2-march-april-2019/big-data-and-clinical-genomics/.

Millican, E. A., Lenzini, P. A., Milligan, P. E., Grosso, L., Eby, C., Deych, E., Grice, G., Clohisy, J. C., Barrack, R. L., Burnett, R. S. J., Voora, D., Gatchel, S., Tiemeier, A., & Gage, B. F. (2007). Genetic-based dosing in orthopedic patients beginning warfarin therapy. *Blood, 110*(5), 1511–1515. https://doi.org/10.1182/blood-2007-01-069609

Nicholson, W. T., Formea, C. M., Matey, E. T., Wright, J. A., Giri, J., & Moyer, A. M. (2021). Considerations when applying pharmacogenomics to your practice. *Mayo Clinic Proceedings, 96*(1), 218–230. https://doi.org/10.1016/j.mayocp.2020.03.011

Obeng, A. O., Kaszemacher, T., Abul-Husn, N., Gottesman, O., Vega, A., Waite, E., Myers, K., Cho, J., Bottinger, E., Ellis, S., & Scott, S. (2016). Implementing algorithm-guided warfarin dosing in an ethnically diverse patient population using electronic health records and preemptive CYP2C9 and VKORC1 genetic testing. *Clinical Pharmacology & Therapeutics, 100*(5), 427–430. https://doi.org/10.1002/cpt.425

Oni-Orisan, A., Mavura, Y., Banda, Y., Thornton, T. A., & Sebro, R. (2021). Embracing genetic diversity to improve black health. *New England Journal of Medicine, 384*(12), 1163–1167. https://doi.org/10.1056/NEJMms2031080

Osheroff, J., Information, & Society, M. S. (2009). *Improving medication use and outcomes with clinical decision support: A step-by-step guide*. Healthcare Information and Management Systems Society Mission.

Overhage, J. M., & McCallie, D. (2020). Physician time spent using the electronic health record during outpatient encounters a descriptive study. *Annals of Internal Medicine, 172*(3), 169–174. https://doi.org/10.7326/M18-3684

Petry, N., & Van Heukalom, J. (2020). Recorded Webcast: Pharmacogenomics in the ICU. In B. Boucher, & C. Haas (Eds.), *Critical Care Self-Assessment Program, 2020 Book 2. Issues in Critical Care Practice*. Lenexa, KS: American College of Clinical Pharmacy. (In press).

Provenzani, A., Santeusanio, A., Mathis, E., Notarbartolo, M., Labbozzetta, M., Poma, P., Provenzani, A., Polidori, C., Vizzini, G., Polidori, P., & D'Alessandro, N. (2013). Pharmacogenetic considerations for optimizing tacrolimus dosing in liver and kidney transplant patients. *World Journal of Gastroenterology, 19*(48), 9156–9173. https://doi.org/10.3748/wjg.v19.i48.9156

Rahawi, S., Naik, H., Blake, K. V., Owusu Obeng, A., Wasserman, R. M., Seki, Y., Funanage, V. L., Oishi, K., & Scott, S. A. (2020). Knowledge and attitudes on pharmacogenetics among pediatricians. *Journal of Human Genetics, 65*(5), 437–444. https://doi.org/10.1038/s10038-020-0723-0

Rasmussen, L. V., Smith, M. E., Almaraz, F., Persell, S. D., Rasmussen-Torvik, L. J., Pacheco, J. A., Chisholm, R. L., Christensen, C., Herr, T. M., Wehbe, F. H., & Starren, J. B. (2019). An ancillary genomics system to support the return of pharmacogenomic results. *Journal of the American Medical Informatics Association, 26*(4), 306–310. https://doi.org/10.1093/jamia/ocy187

Rothschild, A. J., Parikh, S. V., Hain, D., Law, R., Thase, M. E., Dunlop, B. W., DeBattista, C., Conway, C. R., Forester, B. P., Shelton, R. C., Macaluso, M., Brown, K., Lewis, D., Gutin, A., Jablonski, M. R., & Greden, J. F. (2021). Clinical validation of combinatorial pharmacogenomic testing and single-gene guidelines in predicting psychotropic medication blood levels and clinical outcomes in patients with depression. *Psychiatry Research, 296*. https://doi.org/10.1016/j.psychres.2020.113649

Scott, S. A., Sangkuhl, K., Stein, C. M., Hulot, J. S., Mega, J. L., Roden, D. M., Klein, T. E., Sabatine, M. S., Johnson, J. A., & Shuldiner, A. R. (2013). Clinical pharmacogenetics implementation consortium guidelines for CYP2C19 genotype and clopidogrel therapy: 2013 update. *Clinical Pharmacology and Therapeutics, 94*(3), 317–323. https://doi.org/10.1038/clpt.2013.105

Shah, R. R., & Smith, R. L. (2015). Addressing phenoconversion: The Achilles' heel of personalized medicine. *British Journal of Clinical Pharmacology, 79*(2), 222–240. https://doi.org/10.1111/bcp.12441

Shugg, T., Pasternak, A. L., London, B., & Luzum, J. A. (2020). Prevalence and types of inconsistencies in clinical pharmacogenetic recommendations among major U.S. sources. *Npj Genomic Medicine, 5*(1). https://doi.org/10.1038/s41525-020-00156-7

Smith, D. M., Namvar, T., Brown, R. P., Springfield, T. B., Peshkin, B. N., Walsh, R. J., Welsh, J. C., Levin, B., Brandt, N., & Swain, S. M. (2020). Assessment of primary care practitioners' attitudes and interest in pharmacogenomic testing. *Pharmacogenomics, 21*(15), 1085–1094. https://doi.org/10.2217/pgs-2020-0064

Stevenson, J. M., Alexander, G. C., Palamuttam, N., & Mehta, H. B. (2021). Projected utility of pharmacogenomic testing among individuals hospitalized with COVID-19: A retrospective multicenter study in the United States. *Clinical and Translational Science, 14*(1), 153–162. https://doi.org/10.1111/cts.12919

Sun, B., Wen, Y. F., Culhane-Pera, K. A., Lo, M., Xiong, T., Lee, K., Peng, K., Thyagarajan, B., Bishop, J. R., Zierhut, H., & Straka, R. J. (2021). Differences in predicted warfarin dosing requirements between Hmong and East Asians using genotype-based dosing algorithms. *Pharmacotherapy, 41*(3), 265–276. https://doi.org/10.1002/phar.2487

Tavares, L. C., Marcatto, L. R., & Santos, P. C. (2018). Genotype-guided warfarin therapy: Current status. *Pharmacogenomics, 19*(7), 667–685. https://doi.org/10.2217/pgs-2017-0207

Winner, J. G., Carhart, J. M., Altar, C. A., Goldfarb, S., Allen, J. D., Lavezzari, G., Parsons, K. K., Marshak, A. G., Garavaglia, S., & Dechairo, B. M. (2015). Combinatorial pharmacogenomic guidance for psychiatric medications reduces overall pharmacy costs in a 1 year prospective evaluation. *Current Medical Research and Opinion, 31*(9), 1633–1643. https://doi.org/10.1185/03007995.2015.1063483

SECTION THREE

# Adoption and delivery of clinical decision support for pharmacogenomic precision medicine

CHAPTER NINE

# Patient-facing clinical decision support for pharmacogenomic precision medicine

**Khoa Nguyen[1], Kristin Wiisanen[2]**
[1]Pharmacotherapy and Translational Research, College of Pharmacy, University of Florida, Gainesville, FL, United States
[2]UF Health Precision Medicine Program, Center for Pharmacogenomics and Precision Medicine, Department of Pharmacotherapy and Translational Research, College of Pharmacy, University of Florida, Gainesville, FL, United States

## Contents



## Introduction

The need for effective patient education strategies in pharmacogenomics is expanding as testing becomes increasingly available to patients

*Clinical Decision Support for Pharmacogenomic Precision Medicine*
ISBN: 978-0-12-824453-1
https://doi.org/10.1016/B978-0-12-824453-1.00003-8

either directly (e.g., direct-to-consumer [DTC] testing) or through a patient's healthcare provider. Although various static patient education resources (e.g., handouts, printed test results) have been utilized to date, challenges exist in delivering comprehensive patient education in pharmacogenomics. Specifically, pharmacogenomic test results include complex genetic terminology and concepts for which the clinical relevance changes as an individual adds or removes medications or as new evidence emerges (Lucassen & Hall, 2012). In addition, pharmacogenomic tests are lifetime results and ideally require communication across multiple systems as results are often stored in the patient's medical record versus their prescription records. This factor places an additional burden on patient education in pharmacogenomics to ensure that patients both understand and are able to maintain and effectively communicate test results to all members of their healthcare team.

Patient-centered and patient-facing clinical decision support (collectively, patient CDS) provide dynamic mechanisms for patient education that can help to address these challenges. Patient CDS for pharmacogenomics should enable patients to understand and implement clinical recommendations that are anticipatory, accessible, and adaptable. For example, pharmacogenomic test results that are made available through an interactive patient portal that can be accessed on demand and updated in real time as new evidence emerges and drug therapy regimens are revised can improve patient engagement, communication, and decrease information asymmetry between patients and diverse members of their healthcare team (Williams et al., 2018).

In this chapter, foundational concepts influencing the development of patient CDS for pharmacogenomics, regulatory considerations, the current landscape, and opportunities and challenges in developing and implementing patient-facing and patient-centered CDS for pharmacogenomics are reviewed.

## Foundational concepts of patient clinical decision support for pharmacogenomics

CDS systems have historically been employed to improve healthcare delivery and clinical decision-making through provider-facing interventions (e.g., active or passive alerts) that provide evidence-based guidance integrated with patient health data at the point of care (Table 9.1). They may be available as computerized alerts and reminders, computerized guidelines,

**Table 9.1** Common characteristics of pharmacogenomics and nonpharmacogenomics patient-facing CDS.

| Pharmacogenomic patient CDS | Nonpharmacogenomic patient CDS |
|---|---|
| Primarily patient-facing CDS: For germline testing, the primary purpose is to present pharmacogenomic test results to patients, educate patients, and estimate risk of efficacy or toxicity with specific medications; for somatic testing in cancer therapy, somatic test results may also be used to predict efficacy of one or more targeted therapies; most are not integrated with patient's personal health or pharmacy records in real time; are ideally dynamic sources of information with a mechanism for ongoing patient engagement to accommodate revised recommendations with emerging evidence and drug therapy changes; most often available as standalone web-based patient portal supported by commercial or direct-to-consumer laboratory; may be integrated into medical records if housed within patient portal of an existing electronic health record (e.g., MyChart in epic) | May be patient-centered or patient-facing CDS: Commonly available as mobile health (mHealth) applications developed to screen for a specific condition or educate and manage a patient's health and lifestyle; often rely on smartphone technology such as camera functionality and integrate with wearable health technology (e.g., fitness watch); more likely to have consumer-friendly features such as simplified log-in access, real-time tracking, and intuitive data display; more likely to have bidirectional communication to enable patients to upload data (e.g., blood glucose) and receive individualized health recommendations through a closed, disease-specific system; most are not integrated with patient's personal health or pharmacy records in real time, although some integration does exist; common examples include tools to manage medications or medication adherence, blood glucose monitoring or HgA1C monitoring, blood pressure monitoring, screening, or mental health symptom tracking |

Patient-facing CDS strategies may differ according to their purpose.

order sets, patient data reports, documentation templates, and clinical workflow tools (Sutton et al., 2020). In contrast, patient CDS is centered on educating patients, communicating health data, and potentially receiving patient-specific monitoring data to improve health and empower patients in the decision-making process (AfHRa, 2020).

A foundational concept of patient CDS is an appreciation that patients themselves are the individuals who are the ultimate decision-makers in their healthcare and in managing their health. As such, any patient CDS system is only as effective as the patient's ability to understand and apply the information to their health. Therefore, before examining patient CDS further, it is essential to review principles and terminology foundational to this area of study.

Although the terms "patient-centered" and "patient-facing" are often used interchangeably in regards to CDS, they have distinctly different meanings. According to the Agency for Healthcare Research and quality, *patient-centered CDS* refers to decision support that facilitates optimal care for a specific patient. Patient-centered CDS can include evidence-based research, patient-generated health data, or other patient-specific factors (e.g., social, environmental), and involves the patient in using these data to support clinical decision-making. For example, an electronic device that gathers and transmits patients' home blood pressure readings to a patient's clinician and delivers evidence-based recommended actions to the patient based on abnormal readings would be considered patient-centered CDS.

Alternatively, *patient-facing CDS* is defined as CDS that involves direct participation by individual patients based on current evidence and/or patient-specific information (AfHRa, 2020). The majority of currently available pharmacogenomic CDS for patients is comprised of patient-facing CDS.

In its simplest form, patient-facing CDS communicates patient-specific information in a static, unidirectional manner (i.e., to the patient), such as with a web-based patient portal that presents pharmacogenomic test results with general information on affected medications. In this type of system, patients may be able to directly enter medical information and/or have additional resources to ask questions or clarify information, but test results are not integrated automatically with their medical or pharmacy records. This type of portal is often encountered with commercial laboratory testing companies or DTC pharmacogenomic testing. An ideal patient-facing CDS system would integrate and communicate bidirectionally with the patient and provider, rely on accurate medical and pharmacy records, and incorporate other relevant clinical information (e.g., concomitant medications, comorbidities) to facilitate shared-decision-making process between the patient and the provider (Sutton et al., 2020). In terms of patient-centered care, shared decision-making is defined as "an approach where clinicians and patients share the best available evidence when faced with the task of making decisions, and where patients are supported to consider options, to achieve informed preferences" (Elwyn et al., 2012).

Within pharmacogenomics, important predictors of the effectiveness of any patient-facing CDS system are an individual's health literacy, genetic literacy, and e-health literacy (Haga et al., 2014). Health literacy encompasses a patient's ability to perform basic reading and numerical tasks in order to function in a healthcare environment. Genetic literacy refers to having sufficient knowledge of genetic principles to allow for informed medical decision-making. Finally, e-health literacy relates to the patient's ability to understand and appreciate health information from electronic sources and apply this knowledge to address or solve a health-related problem (Haga et al., 2014).

Information asymmetry, or the recognition that one party in a conversation or communication encounter is better informed than the other party, is an especially important concept when considering patient-facing CDS. Within pharmacogenomics, healthcare providers involved in genotype-guided drug therapy changes may include the patient, primary care provider, multiple pharmacists (if patient has prescriptions filled at more than one location), a genetic counselor, and laboratory medicine specialist, each of whom have varying levels of knowledge and comfort surrounding pharmacogenomics and genetics. Therefore, information asymmetry exists not just between the patient and their provider, but among multiple healthcare providers at variable and unpredictable levels. Patient-facing CDS systems, therefore, should account for diverse groups of users who may engage with the system in different ways (Stuckey et al., 2015).

## Regulatory issues surrounding patient-facing CDS for pharmacogenomics

Two agencies have primary authority to regulate genetic testing, including pharmacogenomic tests, the Centers for Medicare and Medicaid Services (CMS), and FDA. The CMS regulates clinical genetic testing under the Clinical Laboratory Improvement Amendments (CLIA) Act of 1988, which established quality standards and strengthened Federal oversight of clinical laboratories in order to ensure the accuracy and reliability of patient test results. The FDA, which regulates medical devices, develops CLIA complexity rules and guidance and categorizes tests based on complexity. FDA first gave notice in 2010 that commercially available genetic test kits that utilized at-home saliva collection methods to obtain a DNA sample were considered medical devices, establishing the precedent that pharmacogenomic tests, and therefore patient laboratory reports that accompany such tests, fall under FDA's medical devices category and regulated as such (Drozda et al., 2018). Controversy has persisted over the past decade regarding FDA's regulatory authority of pharmacogenomic testing.

In October 2018, the first direct-to-consumer (DTC) pharmacogenomic test was approved by FDA, bringing a host of additional regulatory challenges and FDA actions. On October 31st, 2018, the FDA warned against the use of genetic tests with claims to predict patient response to a specific medication. In early 2019, the FDA issued a warning letter to Inova Genomics Laboratory for marketing pharmacogenomic tests that have not been reviewed by the FDA and claim to predict patients' clinical responses to specific drugs. In response to the FDA warning letter, Inova opted to discontinue pharmacogenomic testing, while other firms removed specific medications from labeling and patient reports (FDA Issues Warning Letter to Genomics Lab for Illegally Marketing Genetic Test That Claims to Predict Patients' Responses to Specific Medications, 2019).

The key regulatory issue facing patient CDS in pharmacogenomics concerns FDA's regulation of patient information found in pharmacogenomic laboratory reports, with FDA specifically citing regulation of any test that claims to predict patients' drug responses in its warning letter to Inova. Therefore, information contained in patient CDS is subject to FDA regulation. FDA's actions in this area have been inconsistent and sometimes vague and have led to confusion among clinicians, laboratories, CDS software vendors, and third-party vendors about compliance of patient-facing information about pharmacogenomic test results with regulatory requirements (Gibbs et al., n.d.).

A number of grassroots efforts among the pharmacogenomics community have emerged throughout this controversy. The largest of these efforts, led by the American Society of Pharmacovigilance (https://www.stopadr.org/), is currently working to establish a Collaborative Community with FDA to further communication and clarity for all parties. While this regulatory issue remains controversial and will take time to resolve, it is essential to identify and consider current FDA regulations when developing patient CDS for pharmacogenomics.

## Current landscape of patient-facing clinical decision support for pharmacogenomics

The current landscape of patient-facing CDS spans diverse commercial and research efforts, including clinical patient-facing CDS developed by one or more individual health systems, research-based efforts exploring the effectiveness of patient educational materials in the context of a comprehensive pharmacogenomics implementation, and commercial laboratories working with clinicians or directly with consumers to provide pharmacogenomic testing and support. This section reviews published reports of the use of patient-facing CDS in pharmacogenomics in research or clinical settings.

## Vanderbilt University Medical Center PREDICT initiative

Vanderbilt University Medical Center (VUMC) is one of several institutions participating in National Institutes of Health (NIH)-funded initiatives to encourage early adoption of clinical pharmacogenomic testing. The Pharmacogenomic Resources for Enhanced Decisions in Care and Treatment (PREDICT) initiative was implemented in 2010 (Liu et al., 2021). The full project has been previously described and included end-to-end CDS in the Epic system to translate raw pharmacogenomic data into phenotype interpretations and automated clinical alerts and guidance on medication and/or dosage selection (Pulley et al., 2012). VUMC provided patient-facing CDS integrated into the electronic health record (EHR) through the patient portal, MyHeathAtVanderbilt. PREDICT investigators note the importance of the use of standard pharmacogenomic nomenclature, such as that developed by the Clinical Pharmacogenomics Implementation Consortium (CPIC), secure storage of patient data, clinician and patient accessibility of results, and the need for backward-compatible data structures and nomenclature in developing an end-to-end system that includes patient-facing CDS. VUMC implemented the *Genomic Indicators* module in Epic, which allows for storage of genomic data, to make discrete pharmacogenomic test results available to both clinicians and patients (Liu et al., 2021). Fig. 9.1 shows a display of patient-facing pharmacogenomic data in a system employing the Epic *Genomic Indicators* module. VUMC has also developed an open-access, web-based resource, MyDrugGenome.org, as part of the PREDICT initiative to educate patients about how genetics may affect the way medications work and how genetic results can be incorporated into personalized patient care.

## Geisinger Genomic Medicine Institute

Investigators at the Geisinger Genomic Medicine Institute have developed a patient-facing pharmacogenomics report to communicate pharmacogenomic test results to patients, providers, and pharmacists as part of the Patient

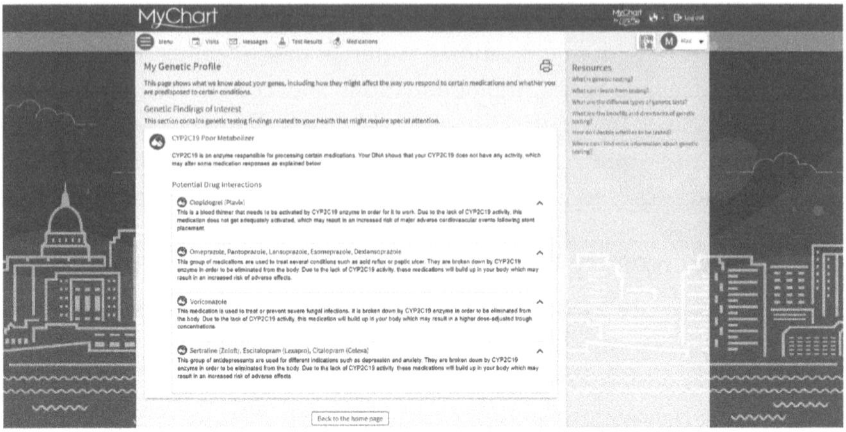


**Figure 9.1** Patient-facing CDS in Epic. Epic allows for displaying pharmacogenomic test results in patient-facing form with current capabilities. *Used with permission from Epic.*

Facing Genomic Test Report Project (Goehringer et al., 2018). This project focused primarily on testing an online, multiuse patient-centered genomic test report that interfaces with the EHR using locally developed software (COMPASS) to manage secure data exchange. The investigation included a pharmacogenomics report that was part of the larger genomics project. Investigators conducted semistructured interviews with ten patients, three primary care providers, three specialists, and four pharmacists to inform development of the COMPASS Pharmacogenomics report. Results of this investigation revealed that patients expressed preference for patient-facing reports that were simple, easy-to-understand, and provided information on how pharmacogenomic test results influenced their drug therapy regimen. Patient participants expressed an expectation that pharmacogenomic test results would be integrated into their EHR and reviewed by their healthcare provider, which supports the use of EHR-integrated patient-facing CDS as compared with external web-based patient portals (Goehringer et al., 2018).

In a companion qualitative analysis conducted to identify patient preferences in patient-facing CDS for genomics, investigators also found that patients specifically stated diagrams are helpful, color is engaging, glossaries and patient-friendly language increased understanding, and categorization of information into discrete sections helps to navigate large amounts of complicated data (Stuckey et al., 2015). Of note, participants in this and other studies specifically requested more concrete information on what actions they should take based on their test results. As previously discussed, this finding creates a particular challenge for the pharmacogenomics industry given FDA regulations regarding the presence of medication-related recommendations in pharmacogenomic test reports.

## UF Health Precision Medicine Program

The University of Florida (UF) Health Precision Medicine Program (PMP), previously referred to as the Personalized Medicine Program, was founded in 2011 (Cavallari et al., 2017). An important focus of the UF Health PMP is on delivering pharmacogenomics to clinical services through electronic CDS, with additional support provided by clinical pharmacists. UF Health utilizes the Epic EHR (without the Epic genomic module, at the time of writing) to build and maintain provider-facing best practice advisories consistent with clinical guidelines and best evidence. UF Health PMP also utilizes the UF Health patient portal, MyUFHealth, to deliver educational video(s) to patients to view prior to their face-to-face appointment in a pharmacogenomics clinical environment supplemented by written patient education materials. Videos are brief (no longer than 2 min) and use short and descriptive animations to explain basic information about pharmacogenomics in patient-friendly terms and notify patients of their eligibility for a pharmacogenomics research study, if applicable. As supported by strategies employed in other implementation sites, hybrid patient education strategies that combine patient-facing CDS, face-to-face patient education, and print materials will likely be required until EHR capabilities for patient-facing education expand.

UF Health PMP also employs a systematic approach that utilizes standardized text constructs to communicate test results via a written pathology lab report and consultation notes, electronic alerts to providers and patients, and written patient materials. To determine which variables influenced standardized language development, UF Health clinicians compared all forms of provider and patient communication of pharmacogenomic test results for multiple gene—drug pairs. It was determined that the language used to convey pharmacogenomic test results within UF Health varied according to organization of the information (e.g., by gene or by indication), its longevity (e.g., recommendations based on current drug therapy or not), how test results are displayed (e.g., genotype vs. phenotype), whether drug therapy recommendations are included, and if phenoconversion was considered in the test result display, if appropriate. Table 9.2 shows an example of such language constructs for a gene—drug pair (Cicali et al., 2020). A systematic approach to CDS and language-development processes can help address the quantity and diversity of factors affecting communication of pharmacogenomic test results.

**Table 9.2** Standardized CDS language for pharmacogenomic return of results according to audience and/or communication strategy.

| Sample affected medication | Test result (laboratory report) | Provider-facing CDS | Pharmacist consultation note | Patient-facing education |
|---|---|---|---|---|
| Selected opioids (e.g., codeine, tramadol) | CYP2D6 UM | "Based on this patient's CYP2D6 genotype, they are at increased risk of adverse events (e.g., respiratory depression, death) with selected opioids, even at low doses. Use non-opioid if possible. If an opioid is indicated, use alternative opioid such as morphine that is not affected by CYP2D6 genotype." | Provides individualized drug therapy recommendations that incorporate other clinical factors (e.g., concomitant medications), and manual phenoconversion calculation (when applicable). | "Based on your CYP2D6 genotype, you have an increased risk of side effects with certain opioids." |

Using standardized language constructs can simplify information delivery based on its purpose and other characteristics.

## eMERGE network

Founded in September 2007, The Electronic Medical Records and Genomics (eMERGE) Network is an NIH-funded. This Network's main goal is to address the implementation of genomic medicine into the US healthcare system. In the third phase of funding (eMERGE3), the Network is developing plans to return results to participants and incorporate genomics results into the EHR (Lynch et al., 2020; Wiesner et al., 2020). There are 10 sites participating, including Cincinnati's Children's Hospital and Medical Center, Children's Hospital of Philadelphia, Columbia University, Geisinger, Kaiser Permanente of Washington/University of Washington, Mayo Clinic, Meharry Medical College, Northwestern University, Partners HealthCare, and Vanderbilt University Medical Center (Fossey et al., 2018; "; Harmonizing Clinical Sequencing and Interpretation for the eMERGE III Network.," 2019). When strategies for returning results were compared among all 10 sites, investigators found that each of the 10 sites had a different process to return results to participants, notify their healthcare providers, and incorporate results into the EHR. The majority of sites (nine institutions) were providing content as printed handouts or pamphlets. Notably, five sites used patient portals or patient health records to deliver content (Rasmussen et al., 2017).

It is important to recognize that printed materials were utilized more widely than electronic delivery systems in many sites of the eMERGE Network. This emphasizes the importance of infrastructure and capability of each system to fully leverage the patient portals or engage patients in different delivery formats. Additionally, adoption of patient portals varies across populations, warranting the need for other modes of delivering patient education (Rasmussen et al., 2017). Finally, projects within the eMERGE Network demonstrated the challenges of finding effective ways to deliver genomic information to either providers or patients. Specifically, the eMERGE experiences reinforces that use of a static version of the pharmacogenomic CDS content introduces a risk of outdated and out-of-sync information being communicated to patients, since genomic knowledge is revolving rapidly. However, the display of dynamic content, utilizing an external, centered content repository, is currently not supported by many EHRs.

## Implementing genomics in pragmatic clinical trials network (IGNITE)

Funded by the National Human Genome Research Institute, the IGNITE Network was established in 2013 to develop methods for effective genomic implementation and incorporation into clinical care. The first phase of the network, IGNITE I, focused on the feasibility of incorporating patient's genomic findings into clinical care while the second phase, IGNITE II, conducted pragmatic clinical trials of genomic medicine interventions. The network is comprised of five multisite clinical groups: University of Florida, Indiana University, Mount Sinai School of Medicine, Duke University, and

Vanderbilt University Medical Center. Many patient-facing clinical decision support and research projects from these sites have been discussed previously in the chapter (Weitzel et al., 2016).

## Ubiquitous PGx (U-PGx) project

Counterpart to the IGNITE and eMERGE Networks in the United States, the Ubiquitous Pharmacogenomics (U-PGx) project was initiated across 7 European countries in 2016 to address the need of pharmacogenomics and CDS implementation (K. Blagec et al., 2018; van der Wouden et al., 2017). The Safety-Code card (Fig. 9.2), perhaps, was the most distinct

**Figure 9.2** Medication safety code card example. In this project, medication safety codes were printed on personalized cards such as this one that patients could carry with them. *Used with permission from the Medication Safety Code Initiative.*

component developed by the U-PGx project to maximize the accessibility and sharing of pharmacogenomic results across healthcare settings, especially in environments that lack an EHR infrastructure (Kathrin Blagec et al., 2016). The safety-code card does not include detailed, patient-specific pharmacogenomic data thus does not need a central storage. After scanning the QR code in the card patients or providers are linked to a website that lists drugs for which important recommendations are available for the card owner (Fig. 9.3).

The Safety-Code card contains an overview of the patient's pharmacogenomic test results and includes a list of relevant drugs affected by pharmacogenomics. Patients who participated in the study were asked to carry this credit card-sized card and provide their test results to their healthcare

**Figure 9.3** User interface for clinical recommendation accompanying medication safety code card. Once the QR code is scanned, clicking on the substance name brings up the corresponding recommendation text. *Used with permission from the Medication Safety Code Initiative.*

provider using the card when needed. Scanning the QR code embedded in the card leads the patients and healthcare professionals to an interactive version of the patient's pharmacogenomic test report (K. Blagec et al., 2018). While the card was developed to allow sharing of pharmacogenomic test results among healthcare settings and healthcare professionals, it is likely that this type of approach can promote patient engagement. Using smartphones or other portable devices, patients can also decode and read the detailed interpretations of their pharmacogenomic test results. The Safety-Code card developed by U-PGx is an appealing solution to solve the challenge of static CDS content.

### Direct-to-consumer pharmacogenomic testing

In most cases, pharmacogenomic testing is ordered by a patient's healthcare provider. The provider determines the necessary pharmacogenomics test(s), orders the test(s) from a commercial or in-house laboratory, obtains test results, determines any needed drug therapy changes, and educates the patient about test results. With DTC pharmacogenomic testing, patients can purchase tests and receive pharmacogenomic and other test results via a secure patient portal, mobile app, and/or in print. While the availability of DTC pharmacogenomic testing may increase patient awareness of and engagement in their test results, it also creates unique challenges for patient CDS. The need for multiple companies to develop and implement proprietary systems for language and format of patient-facing CDS as well as challenges of communicating complex concepts outside of the patient's medical home environment creates an opportunity for inefficiency and risk with DTC patient-facing CDS strategies. In addition, these results should ideally be entered discretely into the patient's electronic pharmacy and medical records and may create provider or institutional liability concerns if patients anticipate a clinical or therapeutic change based on their test results.

## Opportunities and challenges with patient-facing CDS

While the use of patient portals and mobile apps can increase patient access to and awareness of pharmacogenomic testing, these can also lead to misinformation and patient confusion, which can have a negative psychological impact. Unfortunately, in many cases, patients' decisions involving genetic testing are driven more by misunderstanding and anxiety than by values and understanding (Graves et al., 2011).

Patient CDS has had a positive impact on patient care in many areas of healthcare, including reducing readmission, decreasing the length of stay, and improving medication management (Keine et al., 2019). In the same way, patient CDS in pharmacogenomics has the potential to educate patients and improve access to information to support the adoption of pharmacogenomics services. However, a number of challenges and opportunities should be addressed before widespread adoption of patient CDS is a reality in pharmacogenomics.

## Patient health and genomic literacy

Addressing patient health literacy and genetic literacy in pharmacogenomics is crucial to empowering patients for widespread precision medicine adoption. Although the field of pharmacogenomics is relatively new, the public commonly uses related terminology such as "genes" or "genetics." However, many patients who use these terms are unlikely to have a formal science education in human genetics, especially in pharmacogenomics (Lanie et al., 2004). Similarly, available research has shown a lack of general level of knowledge in cancer genetics. This lack of understanding can lead to misunderstanding and anxiety regarding the potential benefits, risks, and limitations of genetic testing and test results (Press et al., 2001). However, many studies have demonstrated that patient education can increase knowledge, decrease anxiety, and improve clinical decision-making. The use of appropriate educational aids (e.g., visual demonstration, personalizing results, simplifying layout, and content.) can significantly help patients understand and retain pharmacogenomics information and help to overcome challenges in patient health and genomic literacy (Blanchette et al., 2014; Gray et al., 2012; Lanie et al., 2004; Rogith et al., 2016).

## Challenges to patient communication

Patient CDS is developed for patients to better understand pharmacogenomics; therefore, it is recommended that the communication of pharmacogenomic test results to patients focus on the interpretation or meaning of the results for patients (e.g., phenotype) instead of the scientific description of results (e.g., genotype). It has also been recommended to avoid stigmatizing terms such as "mutant," "defective," or "abnormal" since these terms can have negative implications (Bates, 2005; Condit et al., 2004; Sidarous & Lamothe, 1995; Veach et al., 2003). Instead, researchers advise use of neutral terms such as "working" or "not working," for example, to describe the function of enzymes. In addition, terms such as "poor metabolizer" can be misinterpreted as automatically meaning a poor treatment outcome and should also be avoided in patient-facing CDS.

While drug metabolism and gene expression are reported in research using probability, this concept is not typically included in results reports for patients. This can also lead to a misconception that if they have a "poor metabolizer" genotype, they are 100% likely to experience an adverse event. Research findings have demonstrated that patients are more likely to overestimate the usefulness of poor metabolizer findings (Brewer et al., 2012). Additionally, the concept of probability can be difficult to comprehend for patients with low numeracy (Gardner et al., 2011; Peters et al., 2011). Therefore, different formats should be explored to convey concepts such as risk, such as one in five patients instead of 20%.

## Patient input to user-interface design

User-centered design (UCD) is an iterative design approach in which users are involved throughout the process of design via a variety of techniques. UCD has been utilized to design pharmacogenomics clinical decision supports support in many healthcare systems (Marcial et al., 2018). With the challenges of patient literacy and scientific data demonstration, UCD has become even more relevant for patient CDS in pharmacogenomics. Table 9.3 describes several key features for patient-centered and patient-facing CDS. Overall, three types of information have been suggested to promote comprehension: (1) description of actual test in simple language, (2) significant of the results for medications that patients are currently taking, and (3) how the results can be important for future medications (Goehringer et al., 2018; Kasper et al., 1992; Marcial et al., 2018; Truong et al., 2020).

## Ethics of privacy, incidental, and secondary findings

In recent years, patients have raised concerns about genetic information privacy (Francis, 2014). Protecting patient privacy is an essential task for healthcare systems. A secure system is needed to produce and deliver patient-facing CDS. In addition, patients should be informed that their pharmacogenomic data are protected to reduce unnecessary concerns. Truong and colleagues recently reported on patient focus group results after constructing a mock pharmacogenomics web portal to deliver pharmacogenomic information and results to patients. Patients expressed a preference for results provided from an understandable and trustable source, with the majority agreeing that they should

**Table 9.3** Features to include in development of patient-facing CDS.

| Research methods | Key feedback from patients | Source(s) |
|---|---|---|
| Interview and focus group | Simple language; visual aids; consistency; logical organization of results; clear and concise content | Stuckey H, Williams JL, Fan AL, et al. Enhancing genomic laboratory reports from the patients' view: A qualitative analysis. Am J Med Genet A 2015; 167A(10):2238–43. Truong TM, Lipschultz E, Schierer E, Danahey K, Ratain MJ, O'Donnell PH. Patient insights on features of an effective pharmacogenomics patient portal. Pharmacogenet Genomics 2020; 30(9):191–200. |
| Semistructured interviews | Handouts or take-home resources to supplement information; simple, jargon-free language; exclusion of technicalities and details | Liang NSY, Adam S, Elliott AM, Siemens A, du Souich C; CAUSES Study; GenCOUNSEL Study, Friedman JM, Birch P. After genomic testing results: Parents' long-term views. J Genet Couns. 2021 Jun 24. |
| Survey using likert-scale questions | Plain and simplified language; personalizing the results; include table and graphs to simplify the layout and content | Olson JE, Rohrer Vitek CR, Bell EJ, McGree ME, Jacobson DJ, St Sauver JL, Caraballo PJ, Griffin JM, Roger VL, Bielinski SJ. Participant-perceived understanding and perspectives on pharmacogenomics: the Mayo Clinic RIGHT protocol (Right Drug, Right Dose, Right Time). Genet Med. 2017 Jul; 19(7):819–825. |

Simplifying language and providing context for information are key elements of patient-facing CDS.

ultimately decide who has access to their pharmacogenomic information. While all participants agreed that their providers should have access to their pharmacogenomic test results, patients' opinions on sharing information with spouses, parents, siblings, or children differed (Truong et al., 2020).

Incidental and secondary findings, defined as results unrelated to the initial indication for genetic testing, have provoked a debate in patient ethics (Saelaert et al., 2018). Whether incidental and secondary findings should or should not be delivered to patients through patient CDS can be a challenge to healthcare systems. Sharing this information can help patients learn risks related to their conditions and support proactive preventive management such as annual screenings and medication management. However, the potential negative psychological impact of unwanted information should also be considered. Patients may have concerns about employment or insurance discrimination as a result of incidental or secondary findings. Gollust and colleagues' study on patient perceptions showed that most patients (90% of patients surveyed) believed they should have a right to access their own genetic information without a gatekeeping medical professional (Gollust et al., 2011). However, having access to incomplete results or without proper consultation can cause anxiety and uncertainty.

## Health and racial disparity

Despite advances in pharmacogenomics and precision medicine, health and racial disparities persist and have a broad impact on public precision health. Health disparities contribute to limited access to pharmacogenomics and precision medicine services among minorities and economically disadvantaged people. In a recent systematic review of patient-facing materials, authors raised a concern that most studies conducted focus on white and educated patients (Wynn et al., 2018). The lack of patient input from lower health literacy groups and minority groups can significantly affect their ability to participate in pharmacogenomic tests and interpret lab results. A survey by Kaphings and colleagues found that lower education levels and identification as part of a racial or ethnic minority group were associated with confusion when interpreting test results (Kaphingst et al., 2012). Overall, a lack of access to information and misinformation resulting from the inability to interpret pharmacogenomics material correctly can deepen the disparity. Patient CDS, when developed, should take into account the input from minority groups carefully and the healthcare system should ensure that minorities will have access to CDS.

## Other considerations for patient education

Patient CDS should incorporate best practices of patient—provider communication, including the need for mutual patient—clinician respect in a supportive environment, harmonized goals, engaging appropriate collaborative partners when needed, and providing accurate information with full transparency (Paget et al., 2011). As with nongenomics patient education, strategies that have been associated with improved retention, understanding, and patient outcomes should be applied when communicating genomic test results to patients in any manner. As such, patient education for genomic testing should be patient-centered and delivered with empathy and respect for patients' potential fears (Friedman et al., 2011; Marcus, 2013). Although studies in this area consistently show that any education at all is superior to no education, certain electronic formats and/or approaches are linked with greater knowledge and understanding. Specifically, ensuring that the educational session is individualized, patient-centered, structured, and interactive (e.g., as with patient-centered CDS), and providing supplemental written information with illustrations or cartoons within narrative text throughout the patient care process are associated with greater effects on patient knowledge, understanding, and retention (Friedman et al., 2011; Marcus, 2013; Trevena, 2014). Therefore given current limitations of EHR systems for patient-facing and patient-centered CDS, a hybrid approach that communicates patient information in multiple formats may be necessary for some time.

## Conclusion

Communication strategies employing patient-centered and patient-facing CDS offer unique solutions to address challenges in educating patients about pharmacogenomics. Although the development of such communication strategies is increasing, limited data are available regarding best practices for development and implementing patient CDS in pharmacogenomics. It is beneficial to review the current clinical and research landscape and explore challenges and opportunities in patient CDS to understand this field.

## References

Agency for Healthcare Research and Quality. (2020). *Patient-centered and patient-facing clinical decision support.*

Bates, B. R. (2005). Public culture and public understanding of genetics: A focus group study. *Public Understanding of Science, 14*(1), 47—65. https://doi.org/10.1177/0963662505048409

Blagec, K., Koopmann, R., Crommentuijn-Van Rhenen, M., Holsappel, I., Van Der Wouden, C. H., Konta, L., Xu, H., Steinberger, D., Just, E., Swen, J. J., Guchelaar, H. J., & Samwald, M. (2018). Implementing pharmacogenomics decision support across seven European countries: The Ubiquitous Pharmacogenomics (U-PGx) project. *Journal of the American Medical Informatics Association, 25*(7), 893–898. https://doi.org/10.1093/jamia/ocy005

Blagec, K., Romagnoli, K. M., Boyce, R. D., & Samwald, M. (2016). Examining perceptions of the usefulness and usability of a mobile-based system for pharmacogenomics clinical decision support: A mixed methods study. *PeerJ*, (2), e1671. https://doi.org/10.7717/peerj.1671

Blanchette, P. S., Spreafico, A., Miller, F. A., Chan, K., Bytautas, J., Kang, S., Bedard, P. L., Eisen, A., Potanina, L., Holland, J., Kamel-Reid, S., McPherson, J. D., Razak, A. R., & Siu, L. L. (2014). Genomic testing in cancer: Patient knowledge, attitudes, and expectations. *Cancer, 120*(19), 3066–3073. https://doi.org/10.1002/cncr.28807

Brewer, N. T., Richman, A. R., DeFrank, J. T., Reyna, V. F., & Carey, L. A. (2012). Improving communication of breast cancer recurrence risk. *Breast Cancer Research and Treatment, 133*(2), 553–561. https://doi.org/10.1007/s10549-011-1791-9

Cavallari, L. H., Weitzel, K. W., Elsey, A. R., Liu, X., Mosley, S. A., Smith, D. M., Staley, B. J., Winterstein, A. G., Mathews, C. A., Franchi, F., Rollini, F., Angiolillo, D. J., Starostik, P., Clare-Salzler, M. J., Nelson, D. R., & Johnson, J. A. (2017). Institutional profile: University of Florida Health Personalized Medicine Program. *Pharmacogenomics, 18*(5), 421–426. https://doi.org/10.2217/pgs-2017-0028

Cicali, E.,C. K., Arwood, M., Staley, Thomas, C., Dalton, Elsey, Cavallari, L., Johnson, J., Cooper-DeHoff, R., & Wiisanen, K. (2020). Development of standardized clinical resources to support implementation of a preemptive pharmacogenomics testing panel. *Clinical Pharmacology & Therapeutics, 107*.

Condit, C. M., Dubriwny, T., Lynch, J., & Parrott, R. (2004). Lay people's understanding of and preference against the word ?mutation? *American Journal of Medical Genetics, 130A*(3), 245–250. https://doi.org/10.1002/ajmg.a.30264

Drozda, K., Pacanowski, M. A., Grimstein, C., & Zineh, I. (2018). Pharmacogenetic labeling of FDA-approved drugs. *JACC: Basic to Translational Science, 3*(4), 545–549. https://doi.org/10.1016/j.jacbts.2018.06.001

Elwyn, G., Frosch, D., Thomson, R., Joseph-Williams, N., Lloyd, A., Kinnersley, P., Cording, E., Tomson, D., Dodd, C., Rollnick, S., Edwards, A., & Barry, M. (2012). Shared decision making: A model for clinical practice. *Journal of General Internal Medicine, 27*(10), 1361–1367. https://doi.org/10.1007/s11606-012-2077-6

*FDA issues warning letter to genomics lab for illegally marketing genetic test that claims to predict patients' responses to specific medications.*(2019). https://www.fda.gov/news-events/press-announcements/fda-issues-warning-letter-genomics-lab-illegally-marketing-genetic-test-claims-predict-patients.

Fossey, R., Kochan, D., Winkler, E., Pacyna, J. E., Olson, J., Thibodeau, S., Connolly, J. J., Harr, M., Behr, M. A., Prows, C. A., Cobb, B., Myers, M. F., Leslie, N. D., Namjou-Khales, B., Rasouly, H. M., Wynn, J., Fedotov, A., Chung, W. K., Gharavi, A., … Kullo, I. J. (2018). Ethical considerations related to return of results from genomic medicine projects: The eMERGE network (phase III) experience. *Journal of Personalized Medicine, 8*(1). https://doi.org/10.3390/jpm8010002

Francis, L. P. (2014). Genomic knowledge sharing: A review of the ethical and legal issues. *Applied and Translational Genomics, 3*(4), 111–115. https://doi.org/10.1016/j.atg.2014.09.003

Friedman, A. J., Cosby, R., Boyko, S., Hatton-Bauer, J., & Turnbull, G. (2011). Effective teaching strategies and methods of delivery for patient education: A systematic review and practice guideline recommendations. *Journal of Cancer Education, 26*(1), 12–21. https://doi.org/10.1007/s13187-010-0183-x

Gardner, P. H., McMillan, B., Raynor, D. K., Woolf, E., & Knapp, P. (2011). The effect of numeracy on the comprehension of information about medicines in users of a patient information website. *Patient Education and Counseling, 83*(3), 398–403. https://doi.org/10.1016/j.pec.2011.05.006

Gibbs, J., & Cato, ME. (n.d.). *FDA once again changes course on LDTs; eases up on pharmacogenomic test restrictions*. FDA Law Blog 2020.

Goehringer, J. M., Bonhag, M. A., Jones, L. K., Schmidlen, T., Schwartz, M., Rahm, A. K., Williams, J. L., & Williams, M. S. (2018). Generation and implementation of a patient-centered and patient-facing genomic test report in the EHR. *eGEMs (Generating Evidence & Methods to Improve Patient Outcomes), 6*(1). https://doi.org/10.5334/egems.256

Gollust, S. E., Gordon, E. S., Zayac, C., Griffin, G., Christman, M. F., Pyeritz, R. E., Wawak, L., & Bernhardt, B. A. (2011). Motivations and perceptions of early adopters of personalized genomics: Perspectives from research participants. *Public Health Genomics, 15*(1), 22–30. https://doi.org/10.1159/000327296

Graves, K. D., Peshkin, B. N., Luta, G., Tuong, W., & Schwartz, M. D. (2011). Interest in genetic testing for modest changes in breast cancer risk: Implications for SNP testing. *Public Health Genomics, 14*(3), 178–189. https://doi.org/10.1159/000324703

Gray, S. W., Hicks-Courant, K., Lathan, C. S., Garraway, L., Park, E. R., & Weeks, J. C. (2012). Attitudes of patients with cancer about personalized medicine and somatic genetic testing. *Journal of Oncology Practice, 8*(6), 329–335. https://doi.org/10.1200/JOP.2012.000626

Haga, S. B., Mills, R., Pollak, K. I., Rehder, C., Buchanan, A. H., Lipkus, I. M., Crow, J. H., & Datto, M. (2014). Developing patient-friendly genetic and genomic test reports: Formats to promote patient engagement and understanding. *Genome Medicine, 6*(7). https://doi.org/10.1186/s13073-014-0058-6

Harmonizing Clinical Sequencing and Interpretation for the eMERGE III Network. (2019). *American Journal of Human Genetics, 105*(3), 588–605. https://doi.org/10.1016/j.ajhg.2019.07.018

Kaphingst, K. A., McBride, C. M., Wade, C., Alford, S. H., Reid, R., Larson, E., Baxevanis, A. D., & Brody, L. C. (2012). Patients understanding of and responses to multiplex genetic susceptibility test results. *Genetics in Medicine, 14*(7), 681–687. https://doi.org/10.1038/gim.2012.22

Kasper, J. F., Mulley, A. G., & Wennberg, J. E. (1992). Developing shared decision-making programs to improve the quality of health care. *Quality Review Bulletin, 18*(6), 183–190. https://doi.org/10.1016/S0097-5990(16)30531-0

Keine, D., Zelek, M., Walker, J. Q., & Sabbagh, M. N. (2019). Polypharmacy in an elderly population: Enhancing medication management through the use of clinical decision support software platforms. *Neurology and Therapy*, 79–94. https://doi.org/10.1007/s40120-019-0131-6

Lanie, A. D., Jayaratne, T. E., Sheldon, J. P., Kardia, S. L. R., Anderson, E. S., Feldbaum, M., & Petty, E. M. (2004). Exploring the public understanding of basic genetic concepts. *Journal of Genetic Counseling, 13*(4), 305–320. https://doi.org/10.1023/B:JOGC.0000035524.66944.6d

Liu, M., Vnencak-Jones, C. L., Roland, B. P., Gatto, C. L., Mathe, J. L., Just, S. L., Peterson, J. F., Van Driest, S. L., & Weitkamp, A. O. (2021). A tutorial for pharmacogenomics implementation through end-to-end clinical decision support based on ten years of experience from PREDICT. *Clinical Pharmacology and Therapeutics, 109*(1), 101–115. https://doi.org/10.1002/cpt.2079

Lucassen, A., & Hall, A. (2012). Consent and confidentiality in clinical genetic practice: Guidance on genetic testing and sharing genetic information. *Clinical Medicine, Journal of the Royal College of Physicians of London, 12*(1), 5–6. https://doi.org/10.7861/clinmedicine.12-1-5

Lynch, J. A., Sharp, R. R., Aufox, S. A., Bland, S. T., Blout, C., Bowen, D. J., Buchanan, A. H., Halverson, C., Harr, M., Hebbring, S. J., Henrikson, N., Hoell, C., Holm, I. A., Jarvik, G., Kullo, I. J., Kochan, D. C., Larson, E. B., Lazzeri, A., Leppig, K. A., … Williams, J. L. (2020). Understanding the return of genomic sequencing results process: Content review of participant summary letters in the eMERGE research network. *Journal of Personalized Medicine, 10*(2). https://doi.org/10.3390/jpm10020038

Marcial, L. H., Richardson, J. E., Lasater, B., Middleton, B., Osheroff, J. A., Kawamoto, K., Ancker, J. S., van Leeuwen, D., Lomotan, E. A., Al-Showk, S., & Blumenfeld, B. H. (2018). The imperative for patient-centered clinical decision support. *eGEMs (Generating Evidence & Methods to Improve Patient Outcomes), 6*(1). https://doi.org/10.5334/egems.259

Marcus, C. (2013). Strategies for improving the quality of verbal patient and family education: A review of the literature and creation of the EDUCATE model. *Health Psychology and Behavioral Medicine*, 482–495. https://doi.org/10.1080/21642850.2014.900450

Paget, L., Han, P., Nedza, S., Kurtz, P., Racine, E., Russell, S., Santa, J., Schumann, M. J., & Von Kohorn, I. (2011). *Patient-clinician communication: Basic principles and expectations.* NAM Perspectives. https://doi.org/10.31478/201106a

Peters, E., Hart, P. S., & Fraenkel, L. (2011). Informing patients: The influence of numeracy, framing, and format of side effect information on risk perceptions. *Medical Decision Making, 31*(3), 432–436. https://doi.org/10.1177/0272989X10391672

Press, N. A., Yasui, Y., Reynolds, S., Durfy, S. J., & Burke, W. (2001). Women's interest in genetic testing for breast cancer susceptibility may be based on unrealistic expectations. *American Journal of Medical Genetics, 99*(2), 99–110. https://doi.org/10.1002/1096-8628(2000)9999:999<00::AID-AJMG1142>3.0.CO;2-I

Pulley, J. M., Denny, J. C., Peterson, J. F., Bernard, G. R., Vnencak-Jones, C. L., Ramirez, A. H., Delaney, J. T., Bowton, E., Brothers, K., Johnson, K., Crawford, D. C., Schildcrout, J., Masys, D. R., Dilks, H. H., Wilke, R. A., Clayton, E. W., Shultz, E., Laposata, M., McPherson, J., … Roden, D. M. (2012). Operational implementation of prospective genotyping for personalized medicine: The design of the vanderbilt PREDICT project. *Clinical Pharmacology and Therapeutics, 92*(1), 87–95. https://doi.org/10.1038/clpt.2011.371

Rasmussen, L., Overby, C., Connolly, J., Chute, C., Denny, J., Freimuth, R., Hartzler, A., Holm, I., Manzi, S., Pathak, J., Peissig, P., Smith, M., Williams, M., Shirts, B., Stoffel, E., Tarczy-Hornoch, P., Vitek, C. R., Wolf, W., & Starren, J. (2017). Practical considerations for implementing genomic information resources. *Applied Clinical Informatics,* 7(3), 870–882. https://doi.org/10.4338/aci-2016-04-ra-0060

Rogith, D., Yusuf, R. A., Hovick, S. R., Fellman, B. M., Peterson, S. K., Burton-Chase, A. M., Li, Y., Bernstam, E. V., & Meric-Bernstam, F. (2016). Patient knowledge and information-seeking about personalized cancer therapy. *International Journal of Medical Informatics, 88*, 52–57. https://doi.org/10.1016/j.ijmedinf.2016.01.005

Saelaert, M., Mertes, H., De Baere, E., & Devisch, I. (2018). Incidental or secondary findings: An integrative and patient-inclusive approach to the current debate. *European Journal of Human Genetics, 26*(10), 1424–1431. https://doi.org/10.1038/s41431-018-0200-9

Sidarous, M., & Lamothe, E. (1995). Norms and standards of practice in genetic counselling. *Health Law Journal, 3*, 153–174.

Stuckey, H., Williams, J. L., Fan, A. L., Rahm, A. K., Green, J., Feldman, L., Bonhag, M., Zallen, D. T., Segal, M. M., & Williams, M. S. (2015). Enhancing genomic laboratory reports from the patients' view: A qualitative analysis. *American Journal of Medical Genetics, A, 167*(10), 2238–2243. https://doi.org/10.1002/ajmg.a.37174

Sutton, R. T., Pincock, D., Baumgart, D. C., Sadowski, D. C., Fedorak, R. N., & Kroeker, K. I. (2020). An overview of clinical decision support systems: Benefits, risks, and strategies for success. *Npj Digital Medicine, 3*(1). https://doi.org/10.1038/s41746-020-0221-y

Trevena, L. (2014). Assessing, communicating, and managing risk in general practice. *British Journal of General Practice, 64*(621), 166–167. https://doi.org/10.3399/bjgp14X677725

Truong, T. M., Lipschultz, E., Schierer, E., Danahey, K., Ratain, M. J., & O'Donnell, P. H. (2020). Patient insights on features of an effective pharmacogenomics patient portal. *Pharmacogenetics and Genomics*, 191–200. https://doi.org/10.1097/FPC.0000000000000413

Veach, P., LeRoy, B., & Bartels, D. (2003). *Facilitating the genetic counseling process: A practice manual.*

Weitzel, K. W., Alexander, M., Bernhardt, B. A., Calman, N., Carey, D. J., Cavallari, L. H., Field, J. R., Hauser, D., Junkins, H. A., Levin, P. A., Levy, K., Madden, E. B., Manolio, T. A., Odgis, J., Orlando, L. A., Pyeritz, R., Wu, R. R., Shuldiner, A. R., Bottinger, E. P., … Ginsburg, G. S. (2016). The IGNITE network: A model for genomic medicine implementation and research. *BMC Medical Genomics, 9*(1). https://doi.org/10.1186/s12920-015-0162-5

Wiesner, G. L., Rahm, A. K., Appelbaum, P., Aufox, S., Bland, S. T., Blout, C. L., Christensen, K. D., Chung, W. K., Clayton, E. W., Green, R. C., Harr, M. H., Henrikson, N., Hoell, C., Holm, I. A., Jarvik, G. P., Kullo, I. J., Lammers, P. E., Larson, E. B., Lindor, N. M., … Leppig, K. A. (2020). Returning results in the genomic era: Initial experiences of the emerge network. *Journal of Personalized Medicine, 10*(2). https://doi.org/10.3390/jpm10020030

Williams, J. L., Rahm, A. K., Zallen, D. T., Stuckey, H., Fultz, K., Fan, A. L., Bonhag, M., Feldman, L., Segal, M. M., & Williams, M. S. (2018). Impact of a patient-facing enhanced genomic results report to improve understanding, engagement, and communication. *Journal of Genetic Counseling, 27*(2), 358–369. https://doi.org/10.1007/s10897-017-0176-6

van der Wouden, C. H., Cambon-Thomsen, A., Cecchin, E., Cheung, K. C., Dávila-Fajardo, C. L., Deneer, V. H., Dolžan, V., Ingelman-Sundberg, M., Jönsson, S., Karlsson, M. O., Kriek, M., Mitropoulou, C., Patrinos, G. P., Pirmohamed, M., Samwald, M., Schaeffeler, E., Schwab, M., Steinberger, D., Stingl, J., … Guchelaar, H. J. (2017). Implementing pharmacogenomics in europe: Design and implementation strategy of the ubiquitous pharmacogenomics consortium. *Clinical Pharmacology and Therapeutics, 101*(3), 341–358. https://doi.org/10.1002/cpt.602

Wynn, R. M., Adams, K. T., Kowalski, R. L., Shivega, W. G., Ratwani, R. M., & Miller, K. E. (2018). The patient in precision medicine: A systematic review examining evaluations of patient-facing materials. *Journal of Healthcare Engineering*. https://doi.org/10.1155/2018/9541621

CHAPTER TEN

# Role of the genetic counselor in pharmacogenomic precision medicine

**Everett Lally[1], Mercy Laurino[1]**
[1]Department of Cancer Genetics and Prevention, Seattle Cancer Care Alliance, Seattle, WA, United States

## Contents



## Introduction

Broad implementation of pharmacogenomics raises a variety of patient counseling considerations. Like other types of genetic testing information, pharmacogenomic results may have implications that extend beyond the original intent for testing. Considerations arise in how these results impact the patient medical management, family members, genetic privacy, discrimination, and psychosocial concerns unique to the patient. Although healthcare providers attempt to address these concerns, the increasing availability and complexity of these tests is daunting. This chapter discusses genetic counselors as healthcare professionals equipped and increasily utilized to address counseling considerations in the

*Clinical Decision Support for Pharmacogenomic Precision Medicine*
ISBN: 978-0-12-824453-1
https://doi.org/10.1016/B978-0-12-824453-1.00004-X

pharmacogenomic setting. These principles fundamentally influence the content and delivery method of both provider-facing and patient-facing clinical decision support (CDS) strategies. To be effective, CDS methods should be developed in light of these fundamentals to enable providers to identify and address the often complex questions that may arise in clinical pharmacogenomics.

## Genetic counseling in pharmacogenomic precision medicine

Pharmacogenomics is defined as the "the study of the role of inheritance in the individual variation in drug response" (Weinshilboum, 2003). While the aim for pharmacogenomics is to improve drug effectiveness and reduce adverse drug reactions, there remains a gap for broad implementaiton. This is attributed to several factors that include payor reimbursement and a limited number of healthcare professionals who are adequately trained in pharmacogenomics. To meet the demand genetic counselors have been proposed and are increasingly being utilized in clinics where pharmacogenomic testing is conducted as as part of clinical practice.

A variety of clinical pharmacogenomic delivery models have been proposed that utilize genetic counselors as part of an interdisciplinary team. One established clinical delivery model involves a collaborative relationship with pharmacists and genetic counselors in a multidisciplinary setting where genetic counselors may be responsible for education and assisting healthcare providers in the communication of the pharmacogenomic test results while aiding reimbursement, while pharmacists may provide therapeutic support, pharmaceutical knowledge, dose adjustment, and drug therapy monitoring (Mills & Haga, 2013). In this model, providers may also consult with a genetic counselor and pharmacist for pharmacogenomic test interpretation and to understand broad clinical implications (Mills & Haga, 2013). Several health systems have established dedicated pharmacogenomics services using similar clinical structures. The Northshore University Health System's multidisciplinary pharcogogenomics team for example included a medical geneticist, pharmacist, a nurse practitioner, and genetic counselors. They determined that there are several factors as being crucial for development and sustainability of this clinic including (1) a billable-service provider, (2) a process for documentation of relevant medication and family histories, (3) personnel with the knowledge required to interpret pharmacogenomic results, (4) personnel to discuss risks, benefits, and limitations of pharmacogenomic testing, and (5) a mechanism for reporting results (Dunnenberger et al., 2016). Additionally, a recent survey to genetic counseling program leaders in both United States and Canada to (Loudon et al., 2021) found that the majority (85%) of survey respondents believed that pharmacogenomics is within the scope of the practice of genetic counselors and should be provided as part of didactic and clinical coursework in training programs.

Laboratory genetic counselors may also play an important role helping clients with test selection, interpretation, and/or patient and provider educational materials in pharmacogenomics particularly as reporting standards evolve over time. Of important note, these laboratories generally rely on genotypying data and report actionable varaints based on various databases including the Clinical Pharmacogenetetics Implementation Consortium (CPIC) and Pharmacogenomics Knowledge Base (PharmGKB) among others (Tafazoli et al., 2021). These variants are often classifed and ranked based on available data including number of pubslihed studies and clinical evidence to support gene-drug interaction. Of variants analyzed as part of pharmacogenomic testing, some laboratories choose to report secondary findings specific to variants while others do not and variants without clear incidental findings may still present challenging clinical correlation.

Pharmacogenomic variants ultimately must be interpreted within the context of other complex factors related to absorption, distribution, and metabolism, and elimination of medications including genetic factors that are yet to be understood or characterized/quantified through testing. For this reason, testing and risk assessment must be personalized by these additional factors which further supports multidisciplinairy involvement (Zierhut et al., 2017).

To illustrate patient counseling aspects of pharmacogenomic testing and genetic counseling considerations, it is worthwhile to consider the following case example as described by Zierhut et al. (2017) in Box 10.1.

## Box 10.1 Case scenario

A 47-year-old woman and mother of two is referred by her primary care provider to a clinical pharmacist for comprehensive medication management visit with a focus on selecting an alternative antidepressant medication after receiving the results of a pharmacogenetic test panel. She was referred for testing by her pain clinic over concern that she was not achieving adequate pain control. She has a complicated medical history including chart diagnoses of fibromyalgia/chronic pain, major depressive disorder, hypertension, type 2 diabetes, insomnia, and new lab results suggesting further work-up for hyperlipidemia. She is single and lives with her children (12-year-old boy and 16-year-old girl). She denies alcohol, tobacco, or illicit drug use. Current health conditions: Fibromyalgia/chronic pain are treated with Gabapentin and oxycodone for which she has inadequte pain control. Major depressive disorder is treated with paroxetine with patial symptom response, however cannot increase dose due to side effects. Select pharmacogenomic results are listed below:

**CYP2D6 *4/*4 poor metabolizer; CYP2C19 *1/*1 Normal metabolizer:** Pharmacogenomic counseling considerations include possible discussion of alternative antidepressants like CYP2C19 substrates (ie citropram, escitolopram, sertraline) and results may explain dose limitations to paroxetine. Genetic counseling considerations include a detailed review of other family members who may be on antidepressants who may be benefit from alternative drug therapies. This discussion may also include topic of genetic testing for family members.

**SLCO1B1 rs4149056 *5/*5 Low function:** Pharmacogenomic counseling considerations include that there is high myopathy risk with this paritcular vairant from some lipid lowering agents if the patient needs these in the future. Genetic counseling considerations include the fact that there are diseases associated with SLCO1B1 that include hyperbilirubinemia, rotor type, digenic and SLCO1B1-related altered drug metabolism. A detailed family history reveals that this patient's son was diagnosed as having jaundice as an infiant and so this result may be related to his finding. There is question of whether family members may should be tested for this variant.

**F5 (Factor V) rs6025 A/A (p.R506Q):** Reduced rate of factor Va deactivation: Pharmacogenomic counseling considerations include increased risk of thrombosis; oral contraceptive or hormone replacement therapy should be avoided in women who have FVL. Genetic counseling considerations include that atient's daughter takes brith control, and so considerations should be made for genetic testing of her daughter and possible discussion of alternative birth control methods.

## Genetic counselor practices and training

Genetic counselors are healthcare professionals with specialized education in medical genetics and psychosocial counseling. The National Society of Genetic Counselors defines genetic counseling as, "the process of helping people understand and adapt to the medical, psychological, and familial implications of the genetic contributions to disease" (Resta et al., 2006). This process integrates

- Interpretation of family and medical histories to assess the chance of disease occurrence or recurrence;
- Education about inheritance, testing, management, prevention, resources, and research; and
- Counseling to promote informed choices and adaptation to the risk or condition

Genetic counseling primarily evolved in tandem with understanding the genetic underpinnings of Huntington's disease, a debilitating heritable progressive brain disorder that causes degenerative of nerve cells in the brain and symptoms and importantly presents after childbearing in the fourth and fifth decade of life. Huntington's disease was the first genetic disorder to be mapped using DNA polymorphism and led to the development of a predictive test for the disorder (Gusella et al., 1983). This predictive test further necessitated access to professionals adequately trained in both medical genetics and psychosocial counseling to meet the unique ethical challenges of genomic medicine.

Since its founding at Sarah Lawrence College in 1969, the genetic counseling profession has expanded globally and is currently estimated to have more than 7000 genetic counselors in at least 28 countries (Abacan et al., 2019; Ormond et al., 2018). Training has historically consisted of a master's degree with both clinical and didactic coursework. Most genetic counselors practice clinically, however others work in laboratory, research, and administrative roles across academia, public health, and healthcare disciplines (Resta et al., 2006). The National Society of Genetic Counselors (NSGC) provides the following scope of practice roles for genetic counselors (National Society of Genetic Counselors: Model Legislative Provisions, n.d.):

- Obtain and evaluate individual, family, and medical histories to determine genetic risk for genetic/medical conditions and diseases in a patient, his/her offspring, and other family members;

- Discuss the features, natural history, means of diagnosis, genetic and environmental factors, and management of risk for genetic/medical condition and diseases;
- Identify, order, and coordinate genetic laboratory tests and other diagnostic studies as appropriate for the genetic assessment;
- Integrate genetic laboratory test results and other diagnostic studies with personal and family medical history to assess and communicate risk factors for genetic/medical conditions and diseases;
- Explain the clinical implications of genetic laboratory tests and other diagnostic studies and their results;
- Evaluate the client's or family's responses to the condition or risk of recurrence and provide client-centered counseling and anticipatory guidance;
- Identify and utilize community resources that provide medical, educational, financial, and psychosocial support and advocacy; and
- Provide written documentation of medical, genetic, and counseling information for families and healthcare professionals.

In practice, clinical genetic counselors typically take a detailed personal and family history, conduct a risk assessment, provide education, and facilitate genetic testing. Throughout this process, genetic counselors elicit the patient's thoughts and feelings about genetic testing to establish informed consent. This therapeutic relationship often involves the discussion of sensitive and personal topics (e.g., a detailed explanation of the inheritance pattern and risk posed for family members which may include facilitating communication and cascade testing of family members). The Reciprocal Engagement Model is used as a framework to establish this patient—provider therapeutic relationship (Biesecker, 2020; Veach et al., 2007). The following principles are applied in this model:

1. Information is key and must be accurate in reflecting cutting-edge genomic science.
2. The relationship is integral and should be psychotherapeutic.
3. Patient autonomy is important, but at times may be challenged in the interest of mitigating health risks.
4. Patients are resilient and capable, but some are vulnerable and need further interventions.
5. Patient's affect reflects only one of a triad of psychological responses to genetic information—cognition, affect, and behavior—all of which are important patient/client experiences to be addressed in genetic counseling (Biesecker, 2020).

In contrast to a clinical setting, genetic counselors also work in various laboratory settings in stewardship roles by aiding clients with selection of appropriate tests, collecting appropriate clinical information for result interpretation, and calling out complex results to providers or patients (Cho & Guy, 2020). Other diverse roles for laboratory genetic counselors may include research and development and technical writing and/or development of edcuational materials and testing algorithms. An important part of laboratory stewardship typically involves a review of incoming orders in order to idenify and correct inappropriate orders that would otherwise be compounded at later stages of the testing process. A comprehensive audit at a large reference laboraotry in the United States demonstrated that genetic counselors in a laboratory setting save money for clients and correct inaccurate genetic test orders (Miller et al., 2014).

Genetic counseling is further positioned to expand into other healthcare disciplines as access is accelerated by improved service delivery models, spurred in part by the global COVID-19 pandemic. Further, it has been shown telehealth and/or telephone genetic counseling is as effective as in-person modalities for promoting knowledge, decision-making, and helping patients to manage their anxiety (Schwartz & Britton, 2011). By utilizing telehealth, genetic counselors may expand into fields like pharmacogenomics with greater ease. Overall, genetic counselors have a proven history of promoting and utilizing precision medicine across a variety of healthcare disciplines.

## Patient counseling considerations and challenges for pharmacogenomic precision medicine

Classification of genetic testing results is complex and evolves over time as new data emerges. Patients are challenged by these types of genetic testing results, implications of these results for their personal health and that of family members, and possible insurance discrimination. Additionally, a lack of diversity within patient populations presents a challenge in both how this genetic data is interpreted and raises issues of access within genomic medicine.

### Classification of genetic variants and implications of results

The American College of Medical Genetics and Genomics (ACMG) has established guidelines for classifying variants based on several lines of evidence including population data, variant type, clinical observations, experimental studies, and indirect and computational modeling (Richards et al., 2015). In recommendations that include input from Association for

Molecular Pathology and the College of American Pathologists, ACMG specifically includes the use of standard terminology of "pathogenic," "likely pathogenic," "uncertain significance," "likely benign," *and* "benign" variants. Given the increasing complexity of high-throughput next-generation sequencing, in combination with other testing methodologies, genetic test result interpretation is performed by a board-certified clinical molecular geneticist or molecular genetic pathologist or equivalent and clinical molecular genetic testing should be performed in a CLIA-approved laboratory (Richards et al., 2015).

The term "mutation" vs variant is often used when describing genetic changes. Specifically, a "*mutation*" means a deviation from a standard genetic sequence (Jarvik & Evans, 2017). A "*mutation*" does not necessarily mean that the deviation is disease-causing. It has become problematic when communicating genetic test results using the word "*mutation*" between providers, patients, family members, and other healthcare professionals involved in caring for the patient given the lack of specificity on how the noted genetic change impacts the clinical phenotype. As such, the word "*variant*" or "allele" is now increasingly in use as the accurate term when referring to genetic changes.

As genetic counselors review personal and family history and conduct informed consent, as part of the pretest counseling session, another objective is to educate patients on the possibility that their results may fall into one of the five variant classifications. Patients often assume that genetic test result comes back as a binary result, a "yes" or a "no." In some cases, this may be true as when there is a known variant in a family, the at-risk family members can choose to be tested specifically for this variant and results may come back either as positive (i.e., confirms it is inherited) or a true negative (i.e., the variant was not identified). Cascade testing of at-risk family members when there's a known familial variant offers them an opportunity to have more certainty in understanding the potential health implications as they proceed with genetic testing. This said, the majority of genetic or genomic tests performed are not a single-site type test but include a wide array of gene panels, whole exomes, and even whole genomes.

Overall, the five gene variant classifications are categorized into three main result themes—positive, negative, or uncertain:

### *Positive result*

A *positive result on genetic testing* denotes that a pathogenic or likely pathogenic variant was identified. This means that the presence of a variant within a gene disrupts protein function and is disease-causing and clinically actionable. For many genetic disorders, manifestation of disease and the impact of positive results greatly depends on the pattern of inheritance. In brief, two of the most common patterns of inheritance are dominant and recessive, although other modes of inheritance also exist including X-linked, Y-linked, mitochondrial, and others. Dominant inheritance corresponds with disease risk accompanying a single pathogenic variant in a single copy of a gene. As such, there is typically a 50% chance of inheriting the pathogenic variant from an affected parent. In contrast, a recessive genetic disorder corresponds with inheriting two pathogenic variants, each on separate chromosomes from both parents. For a recessive genetic disorder, a "carrier" is an individual who inherits a single pathogenic variant on a single copy of the gene and is generally not expected to be at risk of disease. For patients who test positive on germline genetic testing, genetic counselors discuss the associated disease risks, expected inheritance pattern, and possible referrals to other providers. They may also facilitate communication of results and genetic testing in family members. This may include follow-up genetic testing to establish egg-derived or sperm-derived inheritance. Genetic counselors work are trained to help patients cope to the psychosocial issues surrounding these often challenging results.

### Negative result

A *negative result on genetic testing* typically means that no variants were identified that are deemed clinically actionable. It is possible that likely benign or benign variants were noted but these do not impact gene function. This result must be interpreted carefully in the context of personal and family medical history information. For example, if a positive result was expected based on the patient meeting clinical guidelines for genetic testing, and/or suspicion of a rare genetic disorder or based on extensive family history, negative genetic testing may be exceptionally challenging to interpret clinically. A negative result may be misinterpreted as having greatly reducing risk. However, it remains possible there is a gene or combination of genes that are yet to be discovered, or due to heritable or untestable epigenetic factors that may be responsible for a patient's disease. In some cases, it may be of benefit to test other family members to aid in clinical interpretation of a negative test result. A negative result from genetic testing is also routinely reflexed to broader testing involving more genes. However, the likelihood of variants of unknown significance or incidental findings increases significantly when this is done since more genes are being tested.

### Uncertain result

Lastly, an *uncertain result on genetic testing* typically means that a variant of unknown significance (VUS) was identified. These uncertain or inconclusive results are often interpreted as benign/normal variation given the high likelihood of being reclassified as such over time. Thus, it is generally not recommended that these variants be used in clinical decision-making and genetic testing of family members is generally not recommended. Genetic testing laboratories also follow ACMG guidelines for assigning the classification of VUS when there is not enough evidence to fully support a more definitive classification. This said, reference commercial laboratories have propriety information that dictates their internal classification of variants which eventually lead to discordant reporting. To highlight, Pepin et al. reported 41% significant discrepancies when they reviewed variant interpretation previously performed in commercial laboratories (Pepin et al., 2016). Further research may help to clarify the interpretation of these results over time. Many laboratories have mechanisms for communicating VUS reclassifications and will attempt to notify the ordering provider so that the patient's care can be managed accordingly. Some laboratories use online patient portals where updated results may be viewed online directly by the patient. Additionally, for some VUS's which there is high clinical suspicion of pathogenicity, it may be of benefit to test affected family members for personalized risk assessment to aid in reclassification. Many commercial genetic testing laboratories have mechanisms in place for variant reinterpretation and some may include genetic testing of affected family members at no additional cost. Additionally, patients who are found to harbor a VUS in a gene may consider enrolling in a patient registry so that researchers are better able to understand the possible cancer risks associated with this variant over time.

## Concerns for discrimination

One concern in the United States, and albeit globally, is how genetic health information impacts insurance, employment, and other markets, collectively referred to as *genetic discrimination*. Cases of this type of discrimination first arose with the actuarial dilemma of how carriers of sickle cell trait would be accounted for by life insurance and employers. Following the discovery of single gene disorders with well-studied morbidity and mortality, such as Huntington's disease in the 1990s, there was a recognized need for protection against genetic discrimination (Joly et al., 2020). The first federal law in the United States granting some protection for individuals with genetic disorders was the Americans with Disabilities Act (ADA) passed in 1990 (42 U.S.C. §§ 12101–213). Within this legislation, the definition of "physical or mental impairment" importantly was not specified for genetic disorders; however, the law still applied to those individuals who carried a risk for physical or mental impairment but did not display symptoms. Those who met this definition under ADA were protected against discrimination from employment, government services, and public accommodations. In 1996, the Health Insurance Portability and Accountability Act (HIPAA) was passed that included specific language prohibiting the use of genetic information to establish a "preexisting" condition in the absence of a diagnosis and impacted employer-sponsored group plans (42 U.S.C. § 300gg). Years after the passage there was an increasingly recognized need to treat genetic health information with special status, and a unifying law specifically targeting genetic discrimination.

In 2008, the Genetic Information Non-Discrimination Act (GINA) was enacted, which prohibited employers and health insurance companies from discriminating based on "genetic information." This genetic information was more clearly defined as 1) an individual's genetic testing results; 2) genetic tests of family members; and 3) the manifestation of a disease or disorder in family members (42 U.S.C. § 2000ff(4)(A)). Further, "genetic test" referred to an analysis of human DNA, RNA, chromosomes, proteins, or metabolites that detects genotypes, mutations, or chromosomal changes. This definition of genetic testing, however, does not currently include epigenetic information, nor does it include microbiome health information (Joly et al., 2020). Additionally, GINA had limited protections for individuals with preexisting conditions, particularly for individuals who were at risk of a genetic disorder, and then later developed symptoms of that disorder.

In 2008, the Patient Protection and Affordable Care Act (ACA) was also passed, which prohibits discrimination in health insurance of any health-related reason and thus served as a stop gap for limited protections in place from GINA (42U.S.C. § 18001–122). Various state laws were now superseded by GINA and the ACA in granting widespread protections against genetic discrimination in the United States. Other countries have similar models of genetic discrimination protection with Canada, Mexico, and Europe following similar models and protection; however, many countries still lack partial or total protection (Wauters & Van Hoyweghen, 2016).

## Diversity and inclusion in genomic medicine

As part of the National Human Genome Research Institute strategic vision's guiding principles and values for human genomics, Green and colleagues emphasized the importance of striving for "global diversity in all aspects of genomics research, committing to the systematic inclusion of ancestrally diverse and underrepresented individuals in major genomic studies" (Green et al., 2020). This said, a major limitation of pharmacogenomic testing in precision medicine is that majority of published data are informed by research conducted in populations of European descent. As such, there's a gap in fully understanding drug responses within minority populations. Perera et al. emphasized that without increasing pharmacogenomic studies in various minority populations, the overall clinical benefits for pharmacogenetic-guided therapy will not be fruitful. Hence, health disparities will continue and "African Americans and other minority populations will never reach the same playing field" (Perera et al., 2014). To illustrate, a community-based participatory research project was conducted by Holzer et al. in collaboration with the Hmong Genomics Community Advisory Board to better understand their views about personal utility of pharmacogenomic testing and their preferences on how to receive such results (Holzer et al., 2021). Overall, the study participants expressed a positive response to knowing pharmacogenomic test results. The majority felt that the personal and Hmong group results would help their doctors in tailoring medications, if applicable. However, there is concern whether the results can be generalized to the broader Hmong population and how to effectively return such results given its complexity. Simpler write-ups or small group discussions were recommended as opportunities. Furthermore, the limitations of using race and ethnicity in medicine has been ongoingly brought up as a concern. Zhang et al. has reported that the inconsistency of using race and ethnicity categories impacts the reproducibility of pharmacogenomic research and eventually will affect its clinical implications (Zhang & Finkelstein, 2019).

## Conclusion

Genetic counselors play an integral role in precision medicine whose skills and knowledge, in partnership with various healthcare providers, align with disease treatment and prevention. The successful global implementation of precision medicine will require both an increased awareness of the importance of the genetic counselor profession and flexibility in how genetic counselors are incorporated into various healthcare disciplines including pharmacogenomics. It is vitally important that genetic information and personal implications of results be communicated effectively and clearly.

## References

Abacan, M. A., Alsubaie, L., Barlow-Stewart, K., Caanen, B., Cordier, C., Courtney, E., Davoine, E., Edwards, J., Elackatt, N. J., Gardiner, K., Guan, Y., Huang, L. H., Malmgren, C. I., Kejriwal, S., Kim, H. J., Lambert, D., Lantigua-Cruz, P. A., Lee, J. M. H., Lodahl, M., … Wicklund, C. (2019). The global state of the genetic counseling profession. *European Journal of Human Genetics, 27*(2), 183–197. https://doi.org/10.1038/s41431-018-0252-x

Biesecker, B. (2020). Genetic counseling and the central tenets of practice. *Cold Spring Harbor Perspectives in Medicine, 10*(3). https://doi.org/10.1101/cshperspect.a038968

Cho, M. T., & Guy, C. (2020). Evolving roles of genetic counselors in the clinical laboratory. *Cold Spring Harbor Perspectives in Medicine, 10*(10), 1–12. https://doi.org/10.1101/cshperspect.a036574

Drelles, K., Pilarski, R., Manickam, K., Shoben, A. B., & Toland, A. E. (2021). Impact of previous genetic counseling and objective numeracy on accurate interpretation of a pharmacogenetics test report. *Public Health Genomics, 24*(1–2), 26–32. https://doi.org/10.1159/000512476

Dunnenberger, H. M., Biszewski, M., Bell, G. C., Sereika, A., May, H., Johnson, S. G., Hulick, P. J., & Khandekar, J. (2016). Implementation of a multidisciplinary pharmacogenomics clinic in a community health system. *American Journal of Health-System Pharmacy, 73*(23), 1956–1966. https://doi.org/10.2146/ajhp160072

Green, E. D., Gunter, C., Biesecker, L. G., Di Francesco, V., Easter, C. L., Feingold, E. A., Felsenfeld, A. L., Kaufman, D. J., Ostrander, E. A., Pavan, W. J., Phillippy, A. M., Wise, A. L., Dayal, J. G., Kish, B. J., Mandich, A., Wellington, C. R., Wetterstrand, K. A., Bates, S. A., Leja, D., … Manolio, T. A. (2020). Strategic vision for improving human health at the Forefront of Genomics. *Nature, 586*(7831), 683–692. https://doi.org/10.1038/s41586-020-2817-4

Gusella, J. F., Wexler, N. S., Conneally, P. M., Naylor, S. L., Anderson, M. A., Tanzi, R. E., Watkins, P. C., Ottina, K., Wallace, M. R., Sakaguchi, A. Y., Young, A. B., Shoulson, I., Bonilla, E., & Martin, J. B. (1983). A polymorphic DNA marker genetically linked to Huntington's disease. *Nature, 306*(5940), 234–238. https://doi.org/10.1038/306234a0

Holzer, K., Culhane-Pera, K. A., Straka, R. J., Wen, Y. F., Lo, M., Lee, K., Xiong, T., Peng, K., Bishop, J., Thyagarajan, B., & Zierhut, H. A. (2021). Hmong participants' reactions to return of individual and community pharmacogenetic research results: "A positive light for our community". *Journal of Community Genetics, 12*(1), 53–65. https://doi.org/10.1007/s12687-020-00475-3

Jarvik, G. P., & Evans, J. P. (2017). Mastering genomic terminology. *Genetics in Medicine, 19*(5), 491–492. https://doi.org/10.1038/gim.2016.139

Joly, Y., Dupras, C., Pinkesz, M., Tovino, S. A., & Rothstein, M. A. (2020). Looking beyond gina: Policy approaches to address genetic discrimination. *Annual Review of Genomics and Human Genetics, 21*, 491–507. https://doi.org/10.1146/annurev-genom-111119-011436

Loudon, E., Scott, S. A., Rigobello, R., Scott, E. R., Zinberg, R., & Naik, H. (2021). Pharmacogenomic education among genetic counseling training programs in North America. *Journal of Genetic Counseling, 30*(5), 1500–1508. https://doi.org/10.1002/jgc4.1417

Miller, C. E., Krautscheid, P., Baldwin, E. E., Tvrdik, T., Openshaw, A. S., Hart, K., & Lagrave, D. (2014). Genetic counselor review of genetic test orders in a reference laboratory reduces unnecessary testing. *American Journal of Medical Genetics, Part A, 164*(5), 1094–1101. https://doi.org/10.1002/ajmg.a.36453

Mills, R., & Haga, S. B. (2013). Clinical delivery of pharmacogenetic testing services: A proposed partnership between genetic counselors and pharmacists. *Pharmacogenomics, 14*(8), 957–968. https://doi.org/10.2217/pgs.13.76

National Society of Genetic Counselors: Model Legislative Provisions. (n.d.). Retrieved December 20, 2021, from https://www.nsgc.org/Policy-Research-and-Publications/State-Licensure-for-Genetic-Counselors/Model-Legislative-Provisions.

Ormond, K. E., Laurino, M. Y., Barlow-Stewart, K., Wessels, T. M., Macaulay, S., Austin, J., & Middleton, A. (2018). Genetic counseling globally: Where are we now? *American Journal of Medical Genetics, Part C: Seminars in Medical Genetics, 178*(1), 98–107. https://doi.org/10.1002/ajmg.c.31607

Pepin, M. G., Murray, M. L., Bailey, S., Leistritz-Kessler, D., Schwarze, U., & Byers, P. H. (2016). The challenge of comprehensive and consistent sequence variant interpretation between clinical laboratories. *Genetics in Medicine, 18*(1), 20–24. https://doi.org/10.1038/gim.2015.31

Perera, M. A., Cavallari, L. H., & Johnson, J. A. (2014). Warfarin pharmacogenetics: An illustration of the importance of studies in minority populations. *Clinical Pharmacology and Therapeutics, 95*(3), 242–244. https://doi.org/10.1038/clpt.2013.209

Resta, R., Biesecker, B. B., Bennett, R. L., Blum, S., Hahn, S. E., Strecker, M. N., & Williams, J. L. (2006). A new definition of genetic counseling: National Society of Genetic Counselors' Task Force Report. *Journal of Genetic Counseling, 15*(2), 77–83. https://doi.org/10.1007/s10897-005-9014-3

Richards, S., Aziz, N., Bale, S., Bick, D., Das, S., Gastier-Foster, J., Grody, W. W., Hegde, M., Lyon, E., Spector, E., Voelkerding, K., & Rehm, H. L. (2015). Standards and guidelines for the interpretation of sequence variants: A joint consensus recommendation of the American College of Medical Genetics and Genomics and the Association for Molecular Pathology. *Genetics in Medicine, 17*(5), 405–424. https://doi.org/10.1038/gim.2015.30

Schwartz, K. A., & Britton, B. (2011). Use of telehealth to improve chronic disease management. *North Carolina Medical Journal, 72*(3), 216–218.

Tafazoli, A., Wawrusiewicz-Kuurylonek, N., Posmyk, R., & Miltyk, W. (2021). Pharmacogenomics, how to deal with different types of variants in next generation sequencing data in the personalized medicine area. *Journal of Clinical Medicine, 10*(1), 34.

Veach, P. M. C., Bartels, D. M., & LeRoy, B. S. (2007). Coming full circle: A reciprocal-engagement model of genetic counseling practice. *Journal of Genetic Counseling, 16*(6), 713–728. https://doi.org/10.1007/s10897-007-9113-4

Wauters, A., & Van Hoyweghen, I. (2016). Global trends on fears and concerns of genetic discrimination: A systematic literature review. *Journal of Human Genetics, 61*(4), 275–282. https://doi.org/10.1038/jhg.2015.151

Weinshilboum, R. (2003). Inheritance and drug response. *New England Journal of Medicine, 348*(6), 529–537. https://doi.org/10.1056/NEJMra020021

Whirl-Carrillo, M., McDonagh, E. M., Hebert, J. M., Gong, L., Sangkuhl, K., Thorn, C. F., Altman, R. B., & Klein, T. E. (2012). Pharmacogenomics knowledge for personalized medicine. *Clinical Pharmacology and Therapeutics, 92*(4), 414–417. https://doi.org/10.1038/clpt.2012.96

Zhang, F., & Finkelstein, J. (2019). Inconsistency in race and ethnic classification in pharmacogenetics studies and its potential clinical implications. *Pharmacogenomics and Personalized Medicine, 12*, 107–123. https://doi.org/10.2147/PGPM.S207449

Zierhut, H. A., Campbell, C. A., Mitchell, A. G., Lemke, A. A., Mills, R., & Bishop, J. R. (2017). Collaborative counseling considerations for pharmacogenomic tests. *Pharmacotherapy, 37*(9), 990–999. https://doi.org/10.1002/phar.1980

CHAPTER ELEVEN

# A learning health system approach to delivery and adoption of clinical decision support

**Henry Mark Dunnenberger[1], Allen J. Flynn[2]**
[1]Mark R. Neaman Center for Personalized Medicine NorthShore University Health System, Evanston, IL, United States
[2]Department of Learning Health Sciences, School of Information, University of Michigan, Ann Arbor, MI, United States

## Contents



*Clinical Decision Support for Pharmacogenomic Precision Medicine*
ISBN: 978-0-12-824453-1
https://doi.org/10.1016/B978-0-12-824453-1.00008-7

## Introduction

To introduce this chapter, consider what is involved when taking a learning health system (LHS) approach to make improvements. In general, an LHS approach is where a community—organized on a small or large scale—forms and commits to learn and then apply new insights continuously. Within such a *learning community*, ongoing generation of new knowledge happens in the following way. Routine procedures are used to collect and analyze mostly real-world data arising from practice, resulting in new empirical knowledge. When this new knowledge is deemed actionable by members of the community, then it can be implemented as clinical decision support (CDS) or by other means. Once implemented, new empirical knowledge drives practice change toward quality improvement throughout the community and beyond (Friedman et al., 2015, 2017, 2010; Friedman & Rigby, 2013).

In the LHS approach, small-scale learning communities range in size from individual teams to departments and whole institutions. Large-scale communities extend beyond single institutions and may have a regional, statewide, national, or even global scale and reach. This chapter draws its lessons primarily from activity currently taking place at two levels of scale—the smaller institutional level and the larger national level. For the purposes of this chapter, learning by small communities within organizations and nationwide community-based learning are considered.

## The learning cycle

In Fig. 11.1 below, the three-part *learning cycle* portrays how continuous learning generally unfolds in the LHS approach (Friedman et al., 2017). Starting at the bottom of the learning cycle along the course of the green

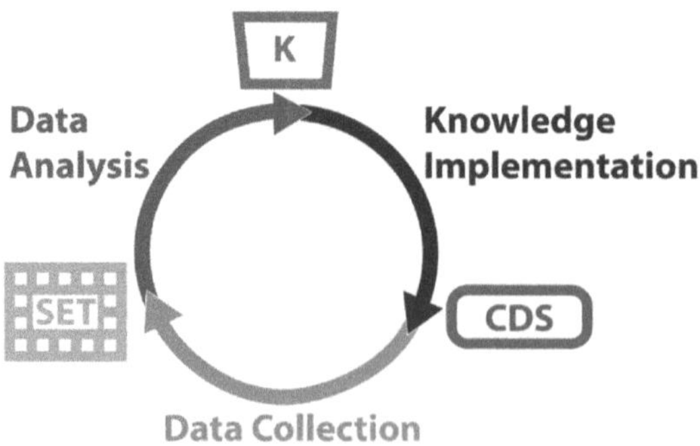


**Figure 11.1** A learning cycle with a data set (SET), knowledge (K), and clinical decision support (CDS).

arrow, the learning community does systematic real-world data collection. Real-world data are a byproduct of actual practice activities, making them different from data collected specifically for randomized clinical trials or other research (Bartlett et al., 2019). Moving around the cycle in the figure, real-world data collection results in a new or updated data set for analysis (SET). Next, on the left-hand side of the learning cycle along the course of the blue arrow, rigorous data analysis takes place. Sometimes, but not always, this data analysis results in new knowledge. New knowledge (K) forms whenever reliable and valid empirical insights are revealed. Then, if the new knowledge gained is actionable, knowledge implementation proceeds on the right-hand side of the cycle. Implementation of new knowledge takes place through a variety of interventions. These interventions support practice change within the learning community and beyond. CDS is one common way of changing practice that is very relevant here (Osheroff et al., 2012). In the learning cycle, the placement of the CDS endpoint where the red arrow returns to the green arrow signifies that, after new CDS has been delivered and adopted, even more real-world data collection can be undertaken to examine the impacts of CDS. It is at this this point that the learning community's operation of a learning cycle can be repeated.

The three fundamental learning cycle processes shown in Fig. 11.1 can be executed to learn about many different types of things. Each time a revolution around the learning cycle finishes, a learning community will have examined some question of interest, learned some new and actionable of value to the community by analyzing real-world data, and implemented that new learning, often by delivering and adopting CDS. Throughout the rest of this chapter, whenever an "LHS approach" is mentioned, it is referring to a community-based approach to operationalize the learning cycle in Fig. 11.1. Next, two examples of complete revolutions around the learning cycle are presented. The first trip around the learning cycle addresses pharmacogenomics and the second trip around the cycle instead addresses the impacts of CDS about pharmacogenomics.

## Operationalizing the LHS approach

### A learning cycle to reveal new pharmacogenomics knowledge

Imagine the following full learning cycle scenario. To answer key pharmacogenomics questions about a group of people and a given drug, a learning community collects genetic, pharmacokinetic, and pharmacodynamic real-world data to create a new data set (Data Collection portion of the learning cycle). Analysis of this new data set has the potential to discover new relationships between one or more genes and responses to the drug (Data Analysis portion of the learning cycle). When novel empirical relationships are established by the learning community in a reliable and valid way—and are deemed actionable in practice—then the learning community has new knowledge to inform and potentially improve the use of the drug. The cycle is completed when the learning community implements its new actionable knowledge by incorporating it into CDS and doing any necessary training and change management activities necessary to evolve clinical practice (Knowledge Implementation portion of the learning cycle).

As more drugs and data from larger, more diverse populations are collected and analyzed, repeated learning cycles generate more actionable pharmacogenomics knowledge, whose implementation becomes increasingly routine throughout the learning community and beyond. Today in the United States, the Clinical Pharmacogenetics Implementation Consortium (CPIC) is an example of a national-scale learning community focused on analyzing previously published scientific results from the field of pharmacogenomics to learn and document what is actionable and advisable in clinical practice (Relling et al., 2020; Relling & Klein, 2011). In the future, organizations such as CPIC will have more and better opportunities to collect and analyze real-world data too.

## A learning cycle to reveal new knowledge about pharmacogenomic clinical decision support

Now imagine a different learning cycle scenario. In this second case, a learning community aims to answer questions not about genes and drugs but instead about the various impacts of its pharmacogenomic CDS. To begin, the learning community collects real-world data about the actual responses to one or more CDS interventions made by a specific group of CDS end-users. The learning community collates these data with other data about related care outcomes to create a new data set. Analysis of this combined data set allows the community to discover new relationships between CDS and care outcomes. When reliable, valid, and actionable relationships between CDS and outcomes are discovered, new empirical knowledge about CDS is established. In cases when this knowledge is deemed actionable by the learning community, it is implemented to improve existing CDS software tools and systems. At that point, this learning cycle focused on improving the design, function, and benefit of pharmacogenomic CDS may repeat again.

## Requirements of the LHS approach

Because of the substantial time and resources required, not all questions can or should be answered using an LHS approach. Based on prior experience operationalizing learning cycles, for the LHS approach to be successful, several questions must be answered in the affirmative to ensure the LHS approach is feasible and appropriate. These questions include the following:

**The Learning Community and its Motivation**

- Is there a learning community that exists at an appropriate level of scale to put the LHS approach into action?

**The Learning Need**

- Is there reason to believe that continuous, ongoing learning over multiple cycles is necessary for the learning community to learn how to advance its practices?

**Data Availability**

- Can data of sufficient quality and quantity be collected or accessed to create appropriate data sets for analysis that will answer questions of interest to the learning community?

**Analytic Methods**

- Are there sufficiently rigorous analytic methods that can be applied to the available data sets to learn something that is new and potentially actionable?

**Ability to Determine if New Knowledge is Actionable**

- Does the learning community have access to sufficient clinical practice expertise to determine whether new knowledge revealed is actionable in practice?

**Ability to Change Practice**

- Can the learning community marshal sufficient human and other resources to implement changes in practice throughout the community?

**Ability to Measure Impacts of Interventions and Assess Progress**

- After changes in practice have been made, are there data collection mechanisms available to provide sufficient additional data to assess the impact of the practice changes?

If a learning community exists that is motivated to solve a problem, and that community can source and analyze sufficient data, learn new things, and apply its new knowledge to the problem in practice, then it may have what it takes to be successful at operationalizing the LHS approach.

# Delivery and adoption of CDS

## Delivery of CDS

In the CDS literature, the term delivery connotes two different things. First, there is the *delivery of technical CDS software logic and tools* through ongoing content updates and new software versions (Khalifa, 2015). This is sometimes also called CDS maintenance. Second, there is the *delivery of CDS advice* to software system end-users, often to end-users of Electronic Health Records (EHRs). This second type of delivery, the delivery of advice generated by CDS for end-users, encompasses how advice is communicated, how its meaning is conveyed, and how the advice is organized, formatted, and displayed alongside other information (Bates et al., 2003). These two different meanings of CDS delivery interact since the delivery of CDS logic and tools consequently enables the delivery of CDS advice. This chapter addresses *both* meanings of CDS delivery in turn.

For operationalizing the LHS approach, it is expected that learning communities formed at various levels of scale will focus on the delivery of potentially better technical CDS software logic and tools and on the delivery of potentially better advice to CDS end-users.

## Adoption of CDS

For this chapter, the term *adoption of CDS* refers to incorporating CDS fully into routine clinical practice. Studies show that end-user engagement with CDS can be idiosyncratic and inconsistent (Cresswell et al., 2017). For that reason, adoption entails committed, consistent use of CDS by individuals and organizations whenever it is relevant in everyday care. An individual or organization that adopts CDS builds it into its workflows, sets rules, and expectations promoting its routine use, and prohibits workarounds that defeat the purposes of CDS.

Scientists who track and study the degree of CDS adoption within organizations draw on many theories and methods. One of these is Rogers' theory of diffusion of innovations. This theory predicts that adoption of new innovations, including CDS, occurs in stages, starting with a tiny number of innovators who do trailblazing work. Per Rogers, from there, adoption grows by including a somewhat larger number of early adopters who others are willing to follow. In an LHS, we anticipate that some innovators and early adopters will be in the membership of a community of interest. Furthermore, we expect that an LHS approach will tend to identify barriers and facilitators to adoption so they can be removed or leveraged, respectively. Engagement of members from groups outside of innovators and early adopters becomes increasingly important as the LHS cycles progress and the barriers and facilitators to adoption change. In an LHS with diverse membership of a community of interest should help ensure adoption across the clinical spectrum.

## An LHS approach to CDS delivery

Taking an LHS approach to learn how to improve both types of CDS delivery may bring benefits that are sufficient to offset the costs of designing, developing, deploying, and maintaining CDS. The rationale for taking an LHS approach to delivery is to discover better ways to deliver technical software logic and tools or better ways to give end-users advice. To identify what better CDS software and advice delivery methods may be, the current methods of CDS delivery must be understood. Next, current CDS delivery methods including technical delivery of software logic and tools and delivery of advice to end-users.

### *Improving the delivery of software logic and tools*

Within institutions, several technical teams that operate installed EHRs must be engaged to deliver relatively modest intra-EHR changes using existing technical capabilities of EHRs. For this reason, delivery of new technical CDS software logic and tools via integration into most modern EHR and health IT environments can be slow and expensive. The technical teams involved have as their members database analysts, application specialists, CDS builders, and quality assurance engineers. Typically, technical CDS changes made within installed EHRs are one-off projects at individual institutions. For many years, health informatics researchers have pointed out the large amount of rework and lack of cross-institutional sharing and consistency of CDS software logic and tool delivery (Hripcsak et al., 1994). To do better, these researchers have explored several other ways of delivering CDS software logic and tools. For example, besides building CDS into EHR directly as intra-EHR CDS, it is now possible to add software apps with new CDS capabilities to existing EHRs as "sidecars" that extend what EHRs can do for end-users (Mandel et al., 2016; Mandl et al., 2012). Open-source and for-profit vendor initiatives to develop and integrate pharmacogenomics CDS apps with EHRs attest to this. Delivery of pharmacogenomics CDS software logic and tools via individual apps offers the potential to achieve greater scale and consistency than the intra-EHR

CDS. On the other hand, use of separate pharmacogenomics CDS apps for technical delivery of software logic and tools has been criticized for its narrowness of scope since there are many other factors that must be considered to provide the highest quality medication therapies.

Since significant inadequacies with technical CDS software logic and tool delivery remain, an LHS approach to finding better, more highly scalable, and more holistic approaches to deliver pharmacogenomics CDS software logic and tools is warranted. An LHS approach may be particularly useful in cases when frequent updates to CDS software logic and tools are necessary, as is currently the case for pharmacogenomics CDS since the pharmacogenomics knowledgebase is evolving very rapidly. Thankfully, informatics researchers, some of whom are inspired by the global success of the World Wide Web as a software delivery platform, are working on new and potentially better ways to deliver software logic and tools more quickly and easily than in the past (Alper et al., 2021).

### Improving the delivery of CDS advice

Several common ways already exist to deliver CDS advice to end-users. One way is to deliver advice using rules that trigger pop-up alerts with messages of advice. CDS alerts can display written advice and images about the genetic determinants of drug selection and dosing. Triggers allow alerts to appear while EHR end users perform certain tasks, like medication prescribing or reviewing medication prescriptions. However, alerts interrupt end-users and sometimes disrupt their train of thought, making them a nuisance and somewhat less effective.

Another common way to deliver pharmacogenomics advice is to embed the advice onscreen in EHR order sets that EHR end-users use to order relevant drugs. One benefit of adding CDS advice to order sets is that it is noninterruptive. However, compared to CDS alerts, it is harder to track and understand end-user responses to noninterruptive CDS advice embedded inside EHR order sets.

Another common way of giving advice to clinicians and patients is to provide patient-specific pharmacogenomics reports. These reports can list many pharmacogenes, related factors, and information items, only some of which are immediately relevant to a given patient's care.

For these common ways of giving CDS advice, determining how best to communicate advice about pharmacogenomics using text and images is also important. Empirical evidence about the best ways to communicate pharmacogenomics advice is lacking.

For alerts and reports, changes to the information content and format are two obvious mechanisms affecting the delivery of CDS advice to end-users. Two other critical aspects of CDS advice delivery are its timing and the degree to which it either disrupts or facilitates the end-user's workflow. Additionally, an unmet challenge of pharmacogenomics CDS advice delivery is to synthesize pharmacogenomic and nonpharmacogenomic factors into more comprehensive, patient-specific medication use advice.

Since these common methods of pharmacogenomics CDS advice delivery—alerts, order sets, and reports—are currently only somewhat effective at ensuring that the genetic determinants of drug selection and dosing are applied accurately and routinely in practice, using an LHS approach to discover better ways to communicate pharmacogenomics advice as CDS are warranted.

Answers to the following questions might be found by taking an LHS approach to the technical delivery of CDS software logic and tools:

- How can the efficiency of technical delivery be measured?
- How can the effectiveness of technical delivery be measured?
- What data standards for inputs and outputs make technical delivery easier?
- How can technical software logic and tools be disseminated rapidly and widely?
- How can software logic errors encountered during technical delivery be minimized?
- What is the greatest degree of technical interoperability that can be achieved for CDS software logic and tools?
- How can technical software logic and tools be consistently kept updated?
- Answers to these other questions might be found by taking an LHS approach to the delivery and communication of advice about pharmacogenomics using CDS:
- What advice content do patients find most useful and why?
- What advice content do prescribers find most useful and why?
- What advice content do pharmacists find most useful and why?
- What advice content do nurses find most useful and why?
- Where is the medication use process do user find advice content most useful and why?
- What are aspects of pharmacogenomics CDS that need to be controlled locally and why?

## An LHS approach to improving the adoption of pharmacogenomic CDS

As expected, there are costs and benefits to taking an LHS approach to CDS adoption. To foster CDS adoption, learning communities rely on policies that pertain at community member institutions and sites. In addition, learning communities facilitate adoption by developing CDS implementation guides to help with deploying pharmacogenomic CDS tools and software and instructional materials to assist with end-user training. The topics of policy support for CDS, technical implementation guides, and instructional materials are addressed in sequence next.

To achieve adoption, it is essential for provider organizations to implement policies that set both the scope and the expectations for routine and consistent pharmacogenomics CDS use in practice and lay out the consequences of not meeting those expectations. Ideally, policies for pharmacogenomics CDS emphasize achieving **mass action** so that the largest possible number of decisions are taken in accordance with the available evidence (Friedman & Flynn, 2019). However, a focus on achieving mass action should not lead to accepting CDS advice uncritically. Although they should be rare, some "exceptions to the rule" must be expected in practice. Policies must allow for these exceptions. In this regard, one reason to take an LHS approach is to systematically gather data on the exceptions that lead pharmacogenomics CDS advice not to be accepted in practice so that the CDS can be continuously improved by accounting for and addressing new exceptions.

Implementation guides facilitate pharmacogenomics CDS adoption. These guides provide checklists and planning documents to guide implementation. An LHS approach could put these implementation guides to test by gathering data about their effectiveness from implementers and using those data to create better implementation guides.

Instructional materials for pharmacogenomics CDS end-users are also important facilitators of adoption. End-users need to know how to use the CDS tool, relevant facts about the underlying science of pharmacogenomics, when they are expected to apply advice from pharmacogenomics CDS in their workflows, and what are the policies governing pharmacogenomics CDS use in their organizations. An LHS approach can help to bring about better instructional materials by gathering data from end-users about the materials and analyzing those data to make improvements to the instructional materials.

Answers to the following list of questions might be found by taking an LHS approach to the adoption of pharmacogenomics CDS:

- What should be the scope of implementation and adoption and why?
- Which policies do and do not stimulate adoption and lead to mass action?
- What makes implementation guides useful to implementers?
- What makes instructional materials useful to end-users?
- What are the determinants of successful adoption?
- How can the use of pharmacogenomics CDS be measured?
- Which tools are actually used in practice? For what purposes? How often?

## The LHS approach to pharmacogenomics CDS in action: a hypothetical case study of Stellar Health

This section reviews a general example that reflects lessons authors have learned about how the LHS approach may play out in a learning community organized inside a single institution. This example is based on an actual case. To make this general example as clear as possible, the work of taking an LHS approach unfolds in a linear, step-by-step fashion. In actuality, the tasks described occurred in parallel and were associated with many fits-and-starts, iterations, and procedural adjustments as they were completed.

### Formation of a learning community

Imagine that a community health system called Stellar Health wants to create a competitive advantage by implementing a system-wide pharmacogenomics program with CDS. Within their large metro region, Stellar Health operates two hospitals, an outpatient surgery center and an oncology center. They also offer extensive primary and specialty care services to outpatients. Will an LHS approach to the delivery and adoption of pharmacogenomics CDS help them?

First, leaders at Stellar Health want to investigate whether they have the capacity to be successful in an LHS approach. To do so, Stellar Health creates a pharmacogenomics subcommittee from its larger pharmacy and therapeutics (P&T) committee. This subcommittee has a diverse membership from across the health system. It includes pharmacists, physicians, administrators, genetic counselors, nurses, system analysts, health information technology personnel, and medical residents. It is determined that Stellar Health does have the capacity—in terms of talent, time, and funding—to move forward using an LHS approach.

The subcommittee's charge is to create a pharmacogenomics program for the delivery and adoption of CDS system-wide over 5 years. Funding is secured to cover genetic testing for up to 2000 patients. Third-party CDS software is sourced, and a 5-year software contract is signed. Leaders of Stellar Health's IT department prioritize the technical work needed. To stay on track, Stellar Health must have a CDS pilot started within 6 months. The project management team is set to report on project metrics on a quarterly basis. Beyond that, the pharmacogenomics subcommittee of the P&T committee is committed to collecting data about the implementation at each step and to documenting lessons learned along with way.

Stellar Health has an integrated EHR which was implemented 15 years ago and has a robust electronic data warehouse with 10 years of high-quality real-world data about patients. Given the goals and timelines of the subcommittee, it is agreed that iterative phases of development and measurement will be needed. Both qualitative and quantitative data are collected throughout the pharmacogenomics CDS implementation. Analyses of these data will drive improvements in the following phases of the pharmacogenomics CDS project.

So far, Stellar Health appears to be well suited to use an LHS approach. They have successfully mobilized capital, leadership, and staff to initiate work. More importantly, they have formed a learning community (the subcommittee and its partners throughout the organization) and the learning community has accepted its mission not only to implement CDS but to continuously learn about pharmacogenomics CDS as they implement it. Next, Stellar Health must tackle the technical delivery of the CDS.

## Technical delivery of pharmacogenomic CDS software logic and tools

The clinicians on the subcommittee want to have CDS that is built into the EHR and accessible at multiple points in the care process. They also would like to have complex, highly targeted alerts which evaluate multiple clinical factors related to drug therapy in addition to genetic factors. However, the IT personnel on the subcommittee propose to use a native EHR-based solution based on discrete genotype-level data. To complicate matters related to technical delivery of CDS, in the early days of the project Stellar Health's executive leadership lets the subcommittee members know that the health system will be acquiring a new hospital in the next 12 months.

Based on previous CDS integrations the team decides key technical delivery metrics will include the amount of technical development and maintenance resources being used to deliver the CDS, the number of CDS report views and the number of CDS alerts triggered per week.

Reviewing the technical specifications of the native EHR tools the subcommittee learns that alerts can only be fired when a medication order is placed and upon pharmacist review of a medication order. While building out the logic for several alerts the IT team notices that using genotype data required by the EHR solution will require them to develop 150 alerts instead of four alerts using phenotype data as found in the CPIC guideline. If genotype data are used, the result would be an extra 300 h of alert development time and 60 h per month extra for maintenance and testing as changes occur in the EHR system. The team also has concerns that the new hospital will not use the same EHR system for the next 5 years. In response to what they have learned thus far, the subcommittee pivots to go with a vendor of a sidecar app solution to provide their pharmacogenomics CDS. The sidecar app can be integrated with various EHRs.

In the sidecar app, CDS alerts can be triggered when patients meet criteria for pharmacogenomic testing. Analysis of Stellar Health data indicates that alerts for testing can help them confirm insurance coverage for testing costs. Upon reviewing the policies of their most common payors, two "coverage likelihood thresholds" are identified, the 75% and 95% likelihood of coverage thresholds. To rule in or rule out coverage at a likelihood of 75%, 10 facts about a patient are needed from Stellar Health's EHR record data. For the 95% threshold, 15 facts are needed. Using historical data to study the implications of these thresholds, the subcommittee determines in the last 6 months 45 patients met the 75% threshold and 43 of these patients met the 95% threshold. Because it takes significantly more compute time to evaluate 15 facts instead of 10 for each patient, the team decides to deploy CDS testing alerts that assess only 10 facts and indicate to end-users that there is a 75% chance genetic testing will be covered for patients identified by the insurance coverage for testing alert. (At this point, the subcommittee has collected and analyzed real-world data and used those data to guide a technical deployment decision. Their decision highlights the truism that more data are not necessarily better. As the subcommittee continues to cycle through the LHS approach, these decisions and implementations will change to meet the evolving needs of the health system.)

After determining how to go about doing their technical delivery of CDS, Stellar Health turns to learning more about the challenges of delivering and communicating advice to end users.

## Delivery of pharmacogenomic CDS advice in practice

Two weeks after turning on the CDS system in the inpatient psychiatric unit, initial reports derived from the system indicate that only pharmacists and physicians are using the pharmacogenomics sidecar app. To collect more data, members of the subcommittee engage in qualitative interviews with clinicians who have interacted with the new sidecar app and its pharmacogenomics CDS.

A few common themes surface from doing user interviews: the alerts are lost in the noise from the other CDS tools in the EHR, the alert text is too long, the alerts fire after the clinician has already made their decision about which medications to use, and there is no way to share the results with the patient when they are discharged. Based on this user feedback, the health IT team at Stellar Health begins making updates to the CDS to adjust the content and layout of the alerts. This work takes about 4 weeks due to competing priorities in the IT department.

The lead nurse on the subcommittee is interested in understanding why no nurse interacted with the app and if this was a gap in the CDS system. She raises this issue during a nursing staff meeting and some nurses comment on how they found the new pharmacogenomics CDS helpful when they are entering phone orders from the physician on call. This news pointed to a problem. The reports being generated about the sidecar app's users inadvertently missed nurse interactions with the sidecar app. Upon review with the pharmacogenomics subcommittee, it is determined the report filtered out nonphysician and nonpharmacist personnel by mistake. This highlights the need for high quality data that are complete in the LHS approach.

Following the real-world clinician interviews, the subcommittee begins reviewing how better technical delivery can make the results and CDS more portable and introducing CDS earlier in the care episode. Meanwhile, a small group of clinicians on the subcommittee decide to engage in community-based research project to understand more about how the Stellar Health's patients want to interact with pharmacogenomic test results and related CDS advice.

With the help of an internal subcommittee functioning as a learning community, Stellar Health has implemented a pharmacogenomics CDS sidecar app alongside their EHRs over about last 9 months. The work to do so included making several changes to the CDS sidecar app and the EHRs. Now they want to analyze and potentially improve their adoption of this pharmacogenomic CDS.

The subcommittee pulls data on number of patients in the inpatient settings with pharmacogenomic testing, number of clinicians interacting with CDS system, number of alerts that triggered, number of times the alerts were overridden, downstream actions after an alert is triggered, and number of pharmacogenomic-based mismatched medications per week.

By the end of the 9 months 75% of the patients seen in the last month had pharmacogenomic results in their medical record. During the first 4 weeks of app use, reports show that only six clinicians interacted with system, representing 30% of the staff. By week 8, an average of 10 clinicians were interacting with the system on a weekly basis. This represented 50% of the staff on any given week.

On average 30 alerts were triggered each week. There was 90% override rate for the alerts. Within 6 h of an alert triggering, the 60% of patients were no longer on the medication that caused the trigger to fire. For patients with pharmacogenomic results the number of genetic mismatched medication dropped from a rate of 20%—5% by the last week. Most of this reduction occurred after the first 6 weeks of the implementation.

Overall, this analysis surfaced several trends of interest. The high override rate coupled with high rate of changes to the trigger medication following an alert suggested that the clinical team may not understand the functionality of the alerts, or the alert functionality could be improved. Adjusting the implementation guide or CDS functionality could improve adoption. The reduction in mismatch medication rates after the first 6 weeks suggests that improvements following the qualitative interviews improved adoption. Finally, since half the clinicians were not interacting with the system 9 months into the project additional mixed method research would be needed to understand how to improve the adoption rate.

While all the metrics are not performing at the top tier, the implementation has clearly been a success. The subcommittee will now take their implementation lessons and strategies from the inpatient psychiatric unit and start to focus on an outpatient primary care clinic. A small group of the subcommittee will stay engaged with the inpatient psychiatric unit to continue to improve their implementation.

This example highlights the many challenges involved in achieving delivery and adoption of pharmacogenomics CDS and how an LHS approach can be used to address them.

## Conclusion

Throughout this chapter, we have tried to communicate the following important features of the LHS approach as it pertains to pharmacogenomics CDS. As the case above highlights, LHS approaches are

- Operationalized by learning communities;
- Applicable at various levels of scale, including the health system level of scale;
- Data-driven and empirical;
- Dependent on having sufficient data, analytic capabilities, practice expertise, and change-making capacity; and
- Suitable for answering certain, but not all, questions about CDS delivery and adoption.

In the future, LHS and other approaches to learning continuously can lead to the wise use of pharmacogenomics CDS that helps optimize medication therapies for patients. Yet, in these early days of both pharmacogenomics CDS and LHSs, it is not clear how best to deliver pharmacogenomics CDS, promote and drive its adoption, or use real-world data to increase its scope and positive impacts. What is clear is that there is a need to learn more about what to do—and what not to do—when it comes to pharmacogenomic CDS.

## References

Alper, B. S., Flynn, A., Bray, B. E., Conte, M. L., Eldredge, C., Gold, S., Greenes, R. A., Haug, P., Jacoby, K., Koru, G., McClay, J., Sainvil, M. L., Sottara, D., Tuttle, M., Visweswaran, S., & Yurk, R. A. (2021). Categorizing metadata to help mobilize computable biomedical knowledge. *Learning Health Systems*. https://doi.org/10.1002/lrh2.10271

Bartlett, V. L., Dhruva, S. S., Shah, N. D., Ryan, P., & Ross, J. S. (2019). Feasibility of using real-world data to replicate clinical trial evidence. *JAMA Network Open, 2*(10). https://doi.org/10.1001/jamanetworkopen.2019.12869

Bates, D. W., Kuperman, G. J., Wang, S., Gandhi, T., Kittler, A., Volk, L., Spurr, C., Khorasani, R., Tanasijevic, M., & Middleton, B. (2003). Ten commandments for effective clinical decision support: Making the practice of evidence-based medicine a reality. *Journal of the American Medical Informatics Association, 10*(6), 523–530. https://doi.org/10.1197/jamia.M1370

Cresswell, K. M., Lee, L., Mozaffar, H., Williams, R., Sheikh, A., Robertson, A., Schofield, J., Coleman, J., Slee, A., Bates, D., Morrison, Z., Girling, A., Chuter, A., Blake, L., Avery, A., Lilford, R., Slight, S., Schofield, B., Shah, S., … McCloughan, L. (2017). Sustained user engagement in health information technology: The long road from implementation to system optimization of computerized physician order entry and clinical decision support systems for prescribing in hospitals in england. *Health Services Research, 52*(5), 1928–1957. https://doi.org/10.1111/1475-6773.12581

Friedman, C. P., & Flynn, A. J. (2019). Computable knowledge: An imperative for learning health systems. *Learning Health Systems, 3*(4). https://doi.org/10.1002/lrh2.10203

Friedman, C., & Rigby, M. (2013). Conceptualising and creating a global learning health system. *International Journal of Medical Informatics, 82*(4), e63–e71. https://doi.org/10.1016/j.ijmedinf.2012.05.010

Friedman, C. P., Rubin, J., Brown, J., Buntin, M., Corn, M., Etheredge, L., Gunter, C., Musen, M., Platt, R., Stead, W., Sullivan, K., & Van Houweling, D. (2015). Toward a science of learning systems: A research agenda for the high-functioning learning health system. *Journal of the American Medical Informatics Association, 22*(1), 43–50. https://doi.org/10.1136/amiajnl-2014-002977

Friedman, C. P., Rubin, J. C., & Sullivan, K. J. (2017). Toward an information infrastructure for global health improvement. *Yearbook of Medical Informatics, 26*(1), 16–23. https://doi.org/10.15265/IY-2017-004

Friedman, C. P., Wong, A. K., & Blumenthal, D. (2010). Achieving a nationwide learning health system. *Science Translational Medicine*, 57cm29. https://doi.org/10.1126/scitranslmed.3001456

Hripcsak, G., Ludemann, P., Pryor, T. A., Wigertz, O. B., & Clayton, P. D. (1994). Rationale for the arden syntax. *Computers and Biomedical Research, 27*(4), 291–324. https://doi.org/10.1006/cbmr.1994.1023

Khalifa, M. (2015). Clinical decision support knowledge management: Strategies for success. In , *Vol. 213. Studies in health technology and informatics* (pp. 67–70). IOS Press. https://doi.org/10.3233/978-1-61499-538-8-67

Mandel, J. C., Kreda, D. A., Mandl, K. D., Kohane, I. S., & Ramoni, R. B. (2016). SMART on FHIR: A standards-based, interoperable apps platform for electronic health records. *Journal of the American Medical Informatics Association, 23*(5), 899–908. https://doi.org/10.1093/jamia/ocv189

Mandl, K. D., Mandel, J. C., Murphy, S. N., Bernstam, E. V., Ramoni, R. L., Kreda, D. A., Michael McCoy, J., Adida, B., & Kohane, I. S. (2012). The SMART platform: Early experience enabling substitutable applications for electronic health records. *Journal of the American Medical Informatics Association, 19*(4), 597–603. https://doi.org/10.1136/amiajnl-2011-000622

Osheroff, J., Teich, J., Levick, D., Saldana, Velasco, F., Sittig, D., Rogers, K., & Jenders, R. A. (2012). *Improving outcomes with clinical decision support: An implementer's guide.*

Relling, M. V., & Klein, T. E. (2011). CPIC: Clinical pharmacogenetics implementation consortium of the pharmacogenomics research network. *Clinical Pharmacology and Therapeutics, 89*(3), 464–467. https://doi.org/10.1038/clpt.2010.279

Relling, M. V., Klein, T. E., Gammal, R. S., Whirl-Carrillo, M., Hoffman, J. M., & Caudle, K. E. (2020). The clinical pharmacogenetics implementation consortium: 10 years later. *Clinical Pharmacology and Therapeutics, 107*(1), 171–175. https://doi.org/10.1002/cpt.1651

CHAPTER TWELVE

# Integration of clinical decision support infrastructure and translating into real-world applications

**Amanda Massmann[1,2], April Schultz[1,2], Joel Van Heukelom[1,2]**
[1]Sanford Imagenetics, Sanford Health, Sioux Falls, SD, United States
[2]Department of Internal Medicine, University of South Dakota Sanford School of Medicine, Sioux Falls, SD, United States

## Contents



## Introduction

This real-world case chapter outlines a clinical scenario in which much of what has been reviewed in this text to this point is put into practice. The objective of this chapter is to describe the innate complexity of clinical decision support (CDS) in a novel area of clinical care and methods to overcome these complexities. This clinical scenario follows patient "DS" through her medical journey and multiple phases of care.

*Clinical Decision Support for Pharmacogenomic Precision Medicine*
ISBN: 978-0-12-824453-1
https://doi.org/10.1016/B978-0-12-824453-1.00007-5

## Challenges with documenting test results in the electronic health record (EHR)

To date, DS has completed various forms of testing independently regarding her ancestry, disease predisposition, and pharmacogenomics through a direct-to-consumer (DTC) testing company (Box 12.1).

### Box 12.1 Patient presentation and initial treatment encounter

DS is a 78-year-old woman with a history of coronary artery disease, hypertension, and dyslipidemia who is actively involved in her healthcare. DS receives care at PrecisionTown Community Hospital, which has recently launched a clinical pharmacogenomics initiative and is in its early stages of implementation.

DS presents to her cardiologist, at PrecisionTown Community Hospital, with chest pressure and shortness of breath. Her cardiologist advises she be taken into the cardiac catheterization laboratory for possible placement of a drug-eluting stent based on her history and presenting symptoms. When the cardiologist places orders for this procedure via the institution's cardiac catheterization orderset, which includes clopidogrel, a pretest alert triggers in the patient's EHR alerting the clinician that a *CYP2C19* order will be placed as part of DS's clinical care.

DS currently does not have genetic results on file; therefore, the clinician explains to DS and her family that genetic testing will be completed in effort to select the most appropriate antiplatelet agent in the event that a stent is placed. DS informs the healthcare team that she has previously had genetic testing and asks her family to obtain her genetic results from her documents at home. The cardiologist explains that the type of DTC genetic testing previously completed was not validated for clinical use, therefore, additional genetic testing is warranted. The clinician orders a single-gene *CYP2C19* test to ensure adequate platelet inhibition with clopidogrel. During her procedure, a stent is placed and afterward, she is discharged on clopidogrel 75 mg once daily and aspirin 81 mg once daily while waiting for results of the *CYP2C19* test.

A few days later her results are returned in the form of a scannable PDF report showing her genotype as CYP2C19 *2/*2, categorized as a CYP2C19 poor metabolizer. The clinician faxes her report to DS's outpatient cardiology clinic. Her cardiologist reviews the results and the Clinical Pharmacogenetics Implementation Consortium (CPIC) guidelines for *CYP2C19*-guided clopidogrel therapy. The cardiologist asks the nursing staff to call and inform DS that she will need to change from clopidogrel to ticagrelor 90 mg twice daily based on her pharmacogenomic test results. Clopidogrel is also added to DS's allergy list to inform future drug therapy choices. Note that DS's discharge paperwork reflects she is taking clopidogrel. The cardiologist sends the PDF of the pharmacogenomic test results to be scanned into the patient's chart in medical records. Each time DS presents to one of her treating clinicians, they will need to access the scanned document to reference these results.

## Limitations of test results provided as printable PDF reports

In many ways the PDF document, a simple hard copy piece of paper, is the most economical and least complicated method for reporting pharmacogenomic test results. Results in PDF format are often made available to a clinician or patient through a vendor-specific portal, via personal email, or faxed copies to healthcare systems. In the age of the EHR, data that are available solely in PDF format can only be incorporated into the medical record by scanning the document into a specified section of the EHR. For these scanned documents to be useful, they need to be sought out within the EHR during every patient visit, or the patient will need to physically bring the document to each appointment. Most providers will likely not realize that pharmacogenomic test results are available within the chart for their use. Of the few who do, many will have limited time to search for the scanned document to access results.

While vendors and healthcare institutions often personalize their pharmacogenomic result PDF documents, the fundamental information needed to make clinical decisions remains consistent. Sections common to most PDF documents include the resulting diplotype/genotype, phenotype, activity score, drug dosing guidance, and laboratory specifications. Dosing guidance is specific to the patient's genetic results and is often displayed in a chart or table. Many vendors use a "stoplight" format (Fig. 12.1) to denote levels of concern among different categories based on published literature, clinical guidelines, and FDA-approved dosing information. The "red" section indicates the highest level of potential clinical impact in which risks

| PHARMACOGENOMIC RESULTS | | |
|---|---|---|
| Patient Name: DS | Date of Birth: 00/00/00 | Medical Record Number: 0000000 |
| **Potentially Impacted Medications** | | |
| **Standard precautions/dosing** | **Moderate interaction/Use with Caution** | **Significant interactions/Consider alternatives** |
| | Omeprazole<br>Lansoprazole<br>Pantoprazole<br>Dexlansoprazole | Clopidogrel<br>Citalopram<br>Escitalopram<br>Sertraline<br>Amitriptyline<br>Clomipramine<br>Doxepin<br>Imipramine<br>Trimipramine<br>Voriconazole |

**Figure 12.1** Sample *CYP2C19* pharmacogenomic test report. Example of pharmacogenomic report utilizing stoplight format (green-yellow-red from left to right). Please note that this report represents single gene testing so is limited in length and details only one gene and diplotype/phenotype recommendation. Multigene pharmacogenomic test panels would provide additional detail of multiple medication classes.

should be weighed with benefits to ensure patient safety and clinical efficacy. Recommendations in the "yellow" may call for a dose adjustment or alternative medication as the potential clinical impact of the drug-gene interaction is less severe. The "green" section indicates that there is no known pharmacogenetic reason to avoid this medication and it can be used as clinically appropriate with standard precautions and dosing recommendations.

Notably, while many patients and clinicians are drawn to the simplicity of the green—yellow-red stoplight format, this approach has been identified as a barrier to implementing pharmacogenomics (Jameson et al., 2021). Patients and providers are at risk of misinterpreting the color coding to mean that medications in the green category are "good" and those in the red category are "bad" and should be avoided at all costs. Instead, the color coding simply informs the severity of the drug-gene interaction. As with any drug therapy decision, clinical, socioeconomic, and other elements are essential factors in the optimal drug therapy decision. Regardless of how the pharmacogenomic test results report is formatted, in practice, it can quickly bring up challenges in implementing a comprehensive genotype-guided approach to patient care delivery (Box 12.2).

**Box 12.2 Patient follow-up encounter in emergency department**

A few weeks later, DS presents to the emergency department, at PrecisionTown Community Hospital, with complaints of shortness of breath, exertional chest pain, and arm numbness. During completion of her history of present illness and medication history, the nurse asks DS to verify her clopidogrel allergy that is listed in the EHR. DS states that she has never had a bad reaction and is not allergic to clopidogrel and the intake nurse deletes the clopidogrel allergy alert.

Clopidogrel 75 mg and aspirin 81 mg are ordered on admission based on DS's previous cardiac catheterization laboratory discharge orders. The following day, the patient's family informs the healthcare team that she should be taking ticagrelor instead of clopidogrel based on her pharmacogenomic testing, although they are unsure of the exact results. The clinicians state they were not aware of this interaction and had no record of any previous reaction to clopidogrel. Upon further examination, clinicians are able to find the scanned PDF report noting that DS is indeed a CYP2C19 poor metabolizer and is at risk of poor response to clopidogrel. DS's antiplatelet therapy is switched to ticagrelor at the previous dose, which is also identified in the outpatient cardiology prescriber's note with some additional searching.

In this example, even though DS had undergone pharmacogenomic testing and the test results were stored in the EHR, they were not readily available or easily accessible to clinicians. As a result, she did not initially receive an optimal drug therapy regimen informed by her genotype. Pharmacogenomic tests resulted in a printable PDF document do not allow for integration into the EHR, preventing the creation of just-in-time guidance in the form of CDS. Information provided in a PDF format cannot be easily updated since they are static, while the information we know about diplotype/genotype to phenotype (and activity scores) translation may change with the availability of additional evidence.

If the allergy list is utilized to record clinical information regarding pharmacogenomic test results, it is important to provide adequate information for clinicians to act on it, which may include an explanation of the genetic test and its clinical relevance for that patient's current and future drug therapy, along with links to supporting evidence, such as published clinical guidelines. In some cases, the EHR has the functionality to link or trigger CDS alerts off of the allergy list, in which case CDS for drug-gene interactions can be built based on this information. Alternatively, if this functionality is not present, the allergy list description will need to supply clinical guidance. This solution is not ideal since it can be cumbersome and inefficient for clinicians and still requires the clinician to seek out the information. No matter what strategy is used, it should be accompanied by education for clinical staff to explain this information, its relevance, how to act upon it clinically, and how to get additional information if needed.

Thus far, this case has illustrated some of the hurdles for successful integration of pharmacogenomic test results into the EHR to guide prescribing for a single drug-gene pair. This process becomes more complicated as a patient's care delivery progresses and their drug therapy regimen is altered over time and there is a need to inform additional providers about the patient's pharmacogenomic test result (Box 12.3).

Pharmacogenomic test results may be manually inputted into patient's charts in multiple formats such as entering information into the allergy or problem list (via SNOMED or ICD-10 codes). Depending on the EHR functionality this can be conducted in a variety of ways, however, any of them can create issues with scalability in the future with interpreting pharmacogenomic test results. While specific to the immediate clinical specialty and scenario, data entry via SNOMED (e.g., "clopidogrel contraindicated" in DS's case) is problematic in that it does not provide any context for this information or other clinical implications. In the current case, relating DS's genotype to a specific drug such as clopidogrel does not provide her PCP with underlying information needed to use genotype information to guide

**Box 12.3 Follow-up encounter with primary care provider**

The inappropriate prescribing of clopidogrel for DS as well as the utilization and incorporation of the allergy list to provide information related to pharmacogenomic testing is discussed at the next PrecisionTown Community Hospital's Medication Safety Committee meeting. The Committee recommends including drug-gene interaction data in the problem list in the EHR instead of the allergy list to decrease the likelihood of human error resulting in inappropriate prescribing. The Committee recommends utilizing ICD-10 codes in the problem list to indicate the presence of a drug-gene interaction. In DS's case, her pharmacogenomic test results are entered into her problem list as "clopidogrel contraindicated" using SNOMED terminology as it was deemed to be more specific by clinicians.

A few weeks after she is discharged from her previous emergency department visit, DS is stable from a cardiology perspective and is responding well to ticagrelor. She is, however, experiencing feelings of hopelessness, worry, and sadness that she just cannot seem to shake and DS makes an appointment to see her primary care provider (PCP), who is within the PrecisionTown Community Hospital outpatient provider network. At her visit with her PCP, DS is diagnosed with major depressive disorder. Upon review of her chart in the EHR, the PCP notices that she has pharmacogenomic test results on file and is a CYP2C19 poor metabolizer. The PCP is unsure how to interpret this information. When she orders citalopram electronically for DS, though, she is not alerted to any concerns with citalopram so she proceeds with prescribing this medication for DS.

When DS returns to her 6-week follow-up appointment, she describes symptoms of nausea and dizziness that she had not felt previously. Her PCP does not find a physiological reason for her symptoms and recommends increasing her dose as she feels it may be due to an anxiety component of DS's depressive disorder. DS agrees and increases her dose with a plan to return in 4 weeks.

DS's antidepressant therapy. Any CDS that could be built based on data entry in this manner would likely trigger off a new order for clopidogrel and therefore is limited in its applicability.

Caution should also be exercised when using ICD-10 codes in the problem list to trigger CDS logic. ICD-10 codes are not specific to each gene and/or phenotype, rendering the codes nonspecific and minimally useful since a poor metabolizer of CYP2C19 has the same ICD-10 code as a poor metabolizer of CYP2D6 (The Web's Free 2021 ICD-10-CM/PCS Medical Coding Reference, n.d.). Since the codes are not specific, it would be difficult or impossible to build CDS for drug-gene interactions linked to ICD-10 codes in the problem list. Alternatively, if the EHR capabilities allow for unique identifiers for each gene and its metabolizer status, this may provide a potential avenue for building CDS.

## Complications with manual entry of discrete laboratory values

Whatever strategy is chosen to enter pharmacogenomic testing results into the EHR, both of the above avenues rely on manual entry of information into the EHR, which is prone to human error (Box 12.4).

### Box 12.4 In search of optimal antidepressant therapy

During the next phase of implementing pharmacogenomic testing at Precision-Town Community Hospital, laboratory components are created allowing for pharmacogenomic test results to be manually entered into the EHR as the initial steps for automated result integration. Discrete laboratory values can be linked to alerts that will trigger when a drug-gene interaction is identified.

During this time, DS begins to experience worsening nausea and dizziness and schedules an appointment with her PCP prior to the planned 4-week routine follow-up. As part of the clinical pharmacogenomics implementation efforts, DS's PCP was recently educated on new functionality within the EHR to help detect potential drug-gene interactions and inform prescribers that they exist. The PCP is not aware of any specific relationship between DS's CYP2C19 poor metabolizer status and her citalopram, but she is suspicious that citalopram may be the cause of DS's symptoms due to their timing and worsening with the increased dose of citalopram. She re-orders citalopram and although an alert does not fire in the EHR, her concern prompts her to obtain a pharmacist consultation for DS.

Upon review of CPIC guidelines, the pharmacist identifies the adverse effects displayed by DS may be attributed to increased blood concentrations of citalopram secondary to CYP2C19 poor metabolizer status. The pharmacist is unsure why an alert did not fire when the prescriber re-ordered citalopram. The pharmacist searches to find the original PDF document that was scanned into the EHR, as well as the genetic results entered manually. The pharmacist discovers an error was made during the manual entry of DS's *CYP2C19* phenotype. DS is a poor metabolizer, but this information was mistakenly entered as normal metabolizer, causing alerts to not be generated for citalopram when it was ordered. The pharmacist collaborates with laboratory and informatics staff to correct the manual entry error and follows up with the PCP to inform her of his findings.

After DS's genetic test results are corrected in the EHR, the prescriber attempts to switch DS from citalopram to escitalopram, which is better tolerated by patients in her experience. However, when entering the order for this medication, a pop-up alert (active interruptive alert) fires to inform the PCP of a drug-gene interaction and prompts the clinician to order a multigene panel that will supply more comprehensive results. The PCP orders a multigene panel for DS.

Entry of pharmacogenomic test results as discrete laboratory values requires a distinct order to be built and available within the EHR for each test with any available resulting values. Examples of the distinct orders that will need to be available include any genes/enzymes that are being tested (*CYP2D6, CYP2C19, SLCO1B1, TPMT, CYP3A5*, etc.). Values available for result entry may include diplotype/genotype, phenotype, and activity scores; all of which can be used to trigger CDS depending on the functionality within the EHR. While it is important that programmed CDS based on manual discrete result entry is possible in many systems, reliance on human factors is time consuming and prone to error. Reliance on the capabilities of a Laboratory Information Management System (LIMS) can provide an alternative (Box 12.5).

**Box 12.5 Patient follow-up encounter with additional pharmacogenomic testing and EHR functionality**

PrecisionTown Community Hospital is in its final phase of clinical pharmacogenomic testing implementation. They are now utilizing a laboratory information management system (LIMS) that sends discrete test result values from DS's multigene panel into the EHR. The primary goal of the LIMS is to handle all the specimens and reagent tracking for all samples going through the molecular genetic laboratory.

DS returns for her follow-up appointment with her PCP once her multigene panel has been completed. Upon ordering medications, her PCP is informed through a pop-up alert (active interruptive alert) that neither citalopram nor escitalopram are preferred antidepressant agents based on DS's pharmacogenomic test results. The alert generates a list of alternative medications based on her pharmacogenomic profile allowing for the option to prescribe an alternative medication if warranted. DS is prescribed a genetically preferred medication and achieves therapeutic efficacy and elimination of adverse effects experienced with previous agents.

Automated result integration is the most advanced, reliable, and beneficial form of pharmacogenomic data representation (Table 12.1). Automated result integration requires the development of laboratory components to house genetic information as discrete values. Results entered as discrete data provide a level of granularity which is both measurable and reportable in a centrally located section of the EHR. It is important to note that integration of genetic results as discrete values is labor intensive and may be cost

**Table 12.1** Summary of pharmacogenomic data interpretation modalities. There are advantages and disadvantages of each reporting and data entry method.

| | Advantages | Disadvantages |
|---|---|---|
| PDF | Inexpensive<br>Easy to create<br>Portable | Data hidden in EHR<br>Difficult to update<br>Difficult to have programmable CDS |
| Manual entry | Programmable CDS is possible<br>Readily available in EHR | Prone to human error<br>Time consuming<br>Difficult to update |
| Automated entry | Programmable CDS is possible<br>Readily available in EHR<br>Real-time<br>Less prone to human error<br>More easily updated | Higher cost<br>Requires significant resources (informatics, pharmacy, bioinformatics, clinicians, etc.) |

prohibitive for some institutions. Native EHR capabilities may allow for more flexible CDS programming, such as providing alternative medications or additional laboratory tests, as compared with some third-party vendors that can provide information and references but do not allow the provider to directly order an alternative medication within the alert.

As knowledge of pharmacogenomics expands and evolves, adaptation of CDS to support genotype-guided prescribing will become increasingly complex (Hoffman et al., 2020). Updating genotype-to-phenotype translation, pharmacogenomic test results, or incorporating activity scores into clinical recommendations can be completed more easily with automated integration as compared with information entered manually into a problem and/or allergy list.

## Applying pharmacogenomic test results throughout a patient's lifetime

Since pharmacogenomic test results are lifetime results, they will impact drug therapy choices throughout a patient's life. In addition, knowledge of a genetic variant may prompt a family member to explore pharmacogenomic testing (Box 12.6).

**Box 12.6 Future drug therapy changes**

Approximately 1 month later, DS calls her PCP to request a refill of her cholesterol medication. Since the multigene panel was recently completed and resulted in the EHR, DS is identified as having a decreased function of SLCO1B1. When the nursing staff enters the order to refill DS's simvastatin, a pop-up alert (active interruptive alert) is triggered and the nurse notifies the prescribing physician. Based on her current dosage and a potential need to increase the dose of simvastatin based on DS's clinical parameters, the PCP elects to see DS in the clinic to discuss her therapeutic options. In a shared decision-making process conducted during her follow-up appointment, it is decided to switch DS to rosuvastatin to decrease the likelihood of DS developing adverse effects.

During her visit, DS's PCP reviews where she can find her pharmacogenomic test results and patient-facing passive CDS within the patient portal of the EHR. DS is delighted to learn she can view medications that are potentially impacted by her pharmacogenomics on her smartphone through an App. DS is instructed that if she needs to seek care outside of the PrecisionTown Community Hospital system, she should show her results to the treatment team. She is also encouraged to utilize this information when discussing any new medications with her treating clinicians. The knowledge DS is able to access within the patient portal allows her to take a more active role in her care and she states it has given her a sense of empowerment.

Many years later, while DS continues to remain active in her own healthcare, family members acknowledge the aging process and are becoming more involved in DS's care. The family identifies the beneficial aspects of incorporating genetics into their loved one's healthcare needs. As a result, DS's children opt to enroll into PrecisionTown Community Hospital's preemptive genetic testing platform to learn more about their personal pharmacogenomic information and how it can impact their care.

## Alerting clinicians to drug-gene interactions after initial pharmacogenomic encounter

There are two main approaches when delivering CDS into the hands of clinicians: active or passive alerts (Wasylewicz & Scheepers-Hoeks, 2018). Active alerts, commonly known as interruptive alerts (Fig. 12.2), leverage technology to drive a clinician to make decisions while interjecting themselves into clinician workflows. Both native EHR and third-party vendors have utilized active interruptive alert functionality. Active

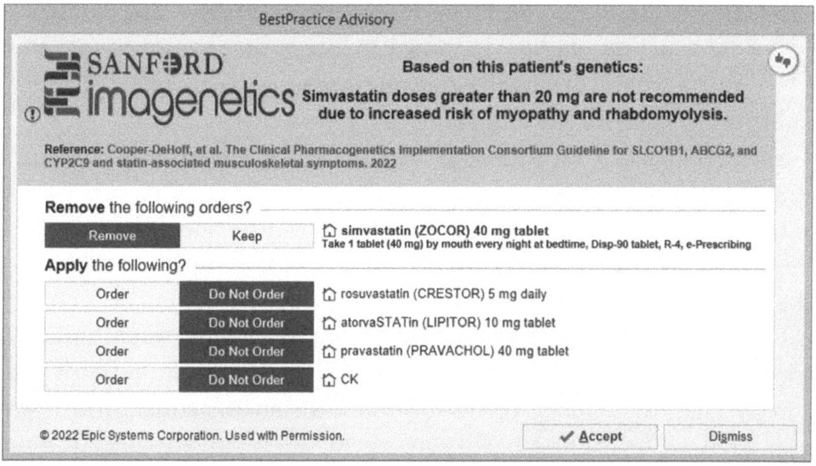


**Figure 12.2** Example of a pop-up alert (active interruptive alert) for simvastatin-***SLCO1B1***. This sample alert is triggered by an order for simvastatin in the presence of an *SLCO1B1* variant. *Used with permission from Epic.*

non-interruptive alerts (inline alerts) are more common in native EHRs and simply provide ordering instructions based on pharmacogenomic test results. Alternatively, passive alerts may require clinicians to take added steps to gather information in the decision-making process. Each individual institution's approach to CDS should be tailored to their needs and capabilities.

## Patient-facing clinical decision support for pharmacogenomic testing

Clinician-facing CDS has been the mainstay of healthcare. With the recent enactment of the 21st Century CURES Act, EHR vendors may leverage CDS to meet to the requirements set forth by the federal government (About ONC's Cures Act Final Rule: Empowering Patients with Their Health Record in a Modern Health IT Economy. The Office of the National Coordinator for Health Information Technology Website, n.d.). Accordingly, access to the totality of an individual's EHR data should be freely available to patients at any given healthcare institution; however, pharmacogenomic information is complex and may warrant discussion with a healthcare team to understand the nuances impacting medication metabolism. The FDA urges pharmacogenomics programs to use caution with the level of information patients are provided regarding medication metabolism as patients may self-adjust medications based on testing results.

Healthcare organizations should be intentional with the mechanism and scope of information sharing in order to meet the requirements of the 21st Century Act and balance these against potential harm when pharmacogenomic test results are not understood.

Genetic results may be accessible through multiple formats depending on the pharmacogenomic testing laboratory. Commercial EHR vendors may have their own platform or an extension that can be added to their EHR to create a secure patient portal (Sutton et al., 2020). Drug-gene interactions can be shared with the patient and allow for additional resources such as patient education materials. The integration of the patient portal within the healthcare institution's EHR allows direct communication with one's healthcare team to facilitate discussion based on pharmacogenomic test results in addition to routine healthcare needs. Alternatively, genetic results and patient facing information may be displayed in a stand-alone platform, such as with DTC testing irrespective of healthcare members' acknowledgment. Healthcare organizations have developed standalone programs to encompass all aspects of this process, from consenting to return of results, in efforts to enhance patient satisfaction.

## Preemptive pharmacogenomic testing for healthy individuals

The turnaround time for pharmacogenomic testing can add logistical constraints if the medication in question must be started immediately, such as with antiplatelet agents initiated post cardiac stent placement. One avenue that many institutions are pursuing is to preemptively provide pharmacogenomic testing to individuals, especially in those who may require genotype-informed drug therapy. The utilization of preemptive testing optimizes the ability to utilize both active and passive CDS modalities to inform drug therapy changes affected by genotype.

# Conclusion

As demonstrated in the patient case example and throughout this chapter, the incorporation of pharmacogenomic test results into clinical care has innate complexities. Utilization of automated data entry and CDS capabilities of the EHR to provide just-in-time guidance to clinicians provides an opportunity to address many of the practical limitations of using pharmacogenomic test results to inform clinical care decisions.

## References

*About ONC's cures act final rule: Empowering patients with their health record in a modern health IT economy. The Office of the National Coordinator for Health Information Technology Website. (n.d.).*

Hoffman, J. M., Flynn, A. J., Juskewitch, J. E., & Freimuth, R. R. (2020). Biomedical data science and informatics challenges to implementing pharmacogenomics with electronic health records. *Annual Review of Biomedical Data Science, 3*(1), 289–314. https://doi.org/10.1146/annurev-biodatasci-020320-093614

Jameson, A., Fylan, B., Bristow, G. C., Sagoo, G. S., Dalton, C., Cardno, A., Sohal, J., & McLean, S. L. (2021). What are the barriers and enablers to the implementation of pharmacogenetic testing in mental health care settings? *Frontiers in Genetics, 12*, 1664–8021. 740216. https://doi.org/10.3389/fgene.2021.740216, 34630531.

Sutton, R. T., Pincock, D., Baumgart, D. C., Sadowski, D. C., Fedorak, R. N., & Kroeker, K. I. (2020). An overview of clinical decision support systems: Benefits, risks, and strategies for success. *Npj Digital Medicine, 3*(1). https://doi.org/10.1038/s41746-020-0221-y

Wasylewicz, A. T. M., & Scheepers-Hoeks, A. M. J. W. (2018). Clinical decision support systems. In *Fundamentals of clinical data science* (pp. 153–169). Springer International Publishing. https://doi.org/10.1007/978-3-319-99713-1_11

*The Web's Free 2021 ICD-10-CM/PCS Medical Coding Reference. (n.d.).*

# Index

'*Note:* Page numbers followed by "f" indicate figures, "t" indicate tables, and "b" indicate boxes.'

## G



## H

## M

## N



## O



## P

## Q



## R



## S



## T

## U



## V



## W

FSC
www.fsc.org
MIX
Papier | Fördert
gute Waldnutzung
FSC® C083411

Kolibri 360 GmbH
Ferdinand-Jühlke-Straße 7
99095 Erfurt, Deutschland
produktsicherheit@kolibri360.de